Nathaniel Edward Yorke-Davies

Foods for the fat

A treatise on corpulency and a dietary for its cure. American Edition

Nathaniel Edward Yorke-Davies

Foods for the fat
A treatise on corpulency and a dietary for its cure. American Edition

ISBN/EAN: 9783337201197

Printed in Europe, USA, Canada, Australia, Japan

Cover: Foto ©berggeist007 / pixelio.de

More available books at **www.hansebooks.com**

FOODS FOR THE FAT:

A TREATISE ON CORPULENCY AND A DIETARY FOR ITS CURE.

BY

NATHANIEL EDWARD DAVIES,

MEMBER OF THE ROYAL COLLEGE OF SURGEONS, ENGLAND.

AMERICAN EDITION.

EDITED BY

CHARLES W. GREENE, M.A., M.D.

PHILADELPHIA:
J. B. LIPPINCOTT COMPANY.
1889.

CONTENTS.

AMERICAN INTRODUCTION.

THAT the reduction of corpulency does not necessarily involve any great hardship to the patient will appear from the study of the following pages. Excessive fatness is to be looked upon as a disease and treated as a disease. It proves, however, when judiciously managed, to be ordinarily a disease not remarkably stubborn or resistful to the means of cure. The systems of diet recommended respectively by Banting, by Oertel, and by Ebstein have all been shown by practical tests to have very considerable objections. The life of the obese man under treatment by either of these systems is likely to become a burden. Deprived to a great degree of some of the chief enjoyments and pleasures of life, his patience is put to a severe test; and very often he is compelled by sheer hunger, and consequent weakness, to forego further treatment, and consequently to go on burdened by his own returning and increasing corpulency.

The present book shows in plain and unscientific language a new and better way. It is perfectly possible, and not very difficult, to go on eating—and eating very well indeed—and yet be cured of excessive stoutness. If the patient be at all well-to-do, he will be able, from these pages, to select for himself a dietary that he and his friends can enjoy together, if need be, while yet he is slowly and safely reducing his surplus of fat.

The author has provided for his patients bills of fare which an epicure need not despise, and which to most readers will seem positively luxurious, but which may be relied upon as certain to reduce corpulency safely, pleasantly, and not too slowly.

With regard to the calendar of foods in their season (pp. 40–45), it must be observed that our American seasons vary greatly from the English, for which reason the English tables here given will hardly serve as a perfect guide to the formation of an American dietary. Our range of market-garden products is much greater than that of the English, and our vegetables (thanks to the Southern market-gardens) are very much longer in season than theirs. On the other hand, there are very few places in the United States which have so rich or so constant a supply of desirable game and fish as that of our British cousins. This calendar is, however, permitted to remain very nearly as in the English edition, for it is full of suggestions which may be turned to good account in this country. Very many of the recipes, pp. 50–132, will be found extremely useful to the householder, apart from their special value in the treatment of obesity. Not a few of these dishes are comparatively unknown in American cookery.

INTRODUCTORY.

WHAT constitutes robust health ? A sound constitution and a lithe, active frame may truly be the answer, for without these the power of enjoying life, even under the most favourable circumstances, is to a great extent limited.

If this be true, there are a large number of people of both sexes whose pleasure is to a considerable extent curtailed—though they may be in other respects fortunate—by the fact that their condition is incompatible with perfect health, and that their bulk renders exercise difficult, and in extreme cases impossible.

The mode of life and dietary of the ordinary Englishman, especially when middle age is reached, undoubtedly tends to foster an accumulation of fat, that, if it does not destroy life directly, does so in very many cases indirectly, by preventing the victim from taking the exercise necessary to circulate the blood, keep the skin acting, and prevent congestion of those internal organs which by their free and healthy action eliminate the waste products of the system.

The disease of Corpulency—for a disease it is—creeps on so insidiously and slowly, and the individual becomes so entangled in its toils, that he or she finds, when it becomes necessary to grapple with it, the power to do so curtailed, and the effort of taking the necessary steps so burdensome as to be practically impossible or too painful to continue.

Happily for such people, science comes to their aid, and, without curtailing very much the pleasures of the table, the diet may be so arranged that, without any danger to health or length of life, a person may slowly and safely reduce bulk and fat to a degree compatible with enjoyment.

How may this be done? A medical adviser who says to the victim of corpulency, 'You must avoid a diet containing sugar and starch'—the principal fatteners—generally gets the reply, 'But, doctor, I don't know what do;' for the ordinary individual does not make dietetics a study, and the reply is perfectly true in the majority of cases. The victim purges and starves himself for a few days, doing serious harm, and making the remedy worse than the disease; or he consults some quack, who reduces his balance at the bank, but not his corporeal redundancy. Under these circumstances, life becomes a burden, and if the victim be well-to-do—and generally he is —he is debarred the pleasures of hunting, fishing, shooting, and all other enjoyable outdoor exercises.

The intention of this little work is to explain as plainly as possible how fat is accumulated in the system, and the foods that form it in excess; and also to give the sufferer a choice of dishes and foods for each meal that will not add to his discomfort, but, if persevered in, will reduce him to proper dimensions. Care has been taken to allow just that amount of food containing fat and starch that is well within the limit of safety.

The task of the dietitian is made easier now a substitute for sugar has been found in saccharin, a harmless product three hundred times sweeter than sugar; and the author, from *personal experiments,* can assure the reader that for all house-hold purposes where it has been necessary to use sugar, such as in tea, coffee, punch, negus, jellies, stewed fruits, apple-sauce, mint-sauce, etc., saccharin is a prefect substitute.

As a dietary that produces fat often produces biliousness and its attendant evils, those who suffer from this ailment may by availing themselves of the foods herein contained avoid much trouble.

The author as a medical man has frequently seen the need of a work of this kind, for diet is as necessary as physic in the treatment of disease, and a dietetic guide relieves the busy practitioner of the tedious process of naming all the articles which the invalid or the sufferer from corpulency may indulge in.

SHERBORNE,
August, 1834.

FOODS FOR THE FAT.

PART I.

CORPULENCY.

1. OF all the evils to which humanity is subject as middle-age creeps on, there is not one more common than excess of fat, or one that causes greater discomfort, or indirectly tends more to shorten life. In men this begins to show itself between the ages of forty and fifty, in women a few years earlier, and though it may not be a disease in itself—unless it attains enormous proportions—it often induces disease by impeding the victim from taking that exercise that nature demands to stimulate the functions of the different organs that keep the body in robust health.

2. About a twentieth part of the weight of the male body should be of fat, and of the female a little more, but it is seldom the balance is so evenly kept. Even where this is greatly exceeded, some people manage to enjoy life and to take a certain amount of exercise, at least in youth; for Daniel Lambert weighed some 450 pounds at the age of twenty-three years, and could then walk from Woolwich to London; subsequently he attained the enormous weight of 739 pounds, and died at the age of thirty-eight years. 'Dr. Warrel records the case of a young married woman, who at eighteen was thin and delicate. She died at the age of fifty-two. The thickness of fat on the chest was four inches, on the abdomen

eight ; her heart, after death, weighed thirty-six ounces (ordinary weight, eight).' Thus in extreme cases one half, or even
four-fifths, of the body may be a mass of fat.

3. Some races of men are more subject to excess of fat than
others, but whether this depends upon heredity or mode of life
is an open question. After the age of forty, particularly in
women, from reasons that may be surmised, excess of fat becomes
almost the rule. This is more common in single females than
in those who have had the care of rearing families. Again,
the Hottentot is almost always protuberant, the German is
proverbially fat, and the Frenchman generally so about the
belly ; the Scotch are thin as a rule ; so are the Irish. What
the Englishman is may be judged by the satire of the age, and
the jolly John Bull sort of man depicted in the pages of *Punch*
may be supposed to represent the national tendency.

4. Among the determining causes of corpulency, the first is,
of course, excess of food, *more especially certain kinds of food*,
and too little work, though some people, curiously enough,
may be very fat and still have poor appetites ; some seem to
get fat, eat what they will,* while others remain thin on the
most luxurious diet. Drink has also its influence. Fat people
usually take a large quantity of liquid, and in some of its
forms, as in sweet wines and malt liquors, it is very fattening.

5. Deficient muscular exercise, by diminishing the amount
of wear of tissue (oxidation of tissue, as physiologists call it),
favours obesity ; and since, as a rule, the stouter the person,
the less capable he is of taking exercise, these two conditions
react one upon the other to the advantage of fat-production.
Nervous influence has much to do with fat ; the high-strung,
nervous individual is seldom obese. On the other hand, the
stupid, heavy, non-intellectual person, or the idiot, is generally
flabby and fat.

6. All those states of the system that prevent the proper
circulation of the blood favour obesity, by limiting its oxy-

* Very *plain* diet may mean *very fattening diet* indeed, as will be seen
further on.

genizing power, by preventing its conversion into carbonic acid and water, and its elimination from the system by the breath. In this way exercise, by rapidly circulating the blood through the lungs, gets rid of fat from the system.*

7. The power of enjoyment is limited in the corpulent person, as exertion is attended with breathlessness, which forbids active exercise. Then, as a matter of course, follows constipation (as the muscular tissue of the bowels gets flabby), and piles and varicose veins come to add to the victim's discomfort. The fat man often ails without apparent cause, is more liable to cold and diarrhœa due to the plethoric and congested state of the mucous membranes, and, dependent on a congested state of the system, to giddiness, headache, flushed face, and a bloated countenance.

8. 'The mental activity,' says Dr. Allchin, ' of the over-fat is variable, and many external causes tend to modify it ; but the temperament is proverbially easy-going, indolent, and lethargic, especially after meals, although very frequently interrupted by attacks of peevishness and irritability, or by unusual somnolence and quiet. Examples, however, of considerable intellectual attainments are not unknown among the corpulent.'

9. The fat man is liable to profuse sweating This, being highly acid, causes chafing in the groins, with painful eruptions. Where this takes place, the parts become sore and inflamed. He is also more subject to gout, and his urine always contains uric acid to excess ; therefore, the same may be said of his liability to rheumatism. He is more liable to disease than a thin person, with this disadvantage, that ailments in him run a more unfavourable course, and he bears treatment worse. Further, he is more difficult to treat on account of his inability to stand lowering measures, and suffers from debility longer during convalescence.

10. When too much food is indulged in, the internal organs

---

* See ' Aids to Long Life,' by the same author, page 193. From that work part of this section has been drawn. The evils of corpulency and other causes of premature decay are treated at length in that book.

become embarrassed by the waste in the system that is not eliminated, and a feeling of weakness ensues, which quickly passes away if exercise is taken and a little abstinence practised. Diet influences the ch racter of men and nations, meat-eating people being more energetic and stronger than those that live mostly on vegetables.† Growth and temperament are also influenced by food, and in the case of bees even the sex.‡

11. With age corpulency increases permanently, unless some exhausting disease such as chronic bronchitis or diabetes come on, so that excessive fat should always be regarded as a grave matter, in every way likely to shorten life, to say nothing of making it a burden by its encumbrance. Death by faintness from an overloaded heart or an overloaded stomach, by apoplexy from congestion and weakness of the blood-vessels, by bronchitis or dropsy from poorness of the blood and the languid state of the circulation, often terminates life about the beginning of the sixth decade.

12. ' Recognising that accumulation of fat is a perversion of nutrition, which, if once established, and with a strong hereditary predisposition, cannot be *cured*, it follows that we should endeavour to prevent as far as possible its increase by avoidance of those factors which science tells us are favourable to its development. The cardinal rule in any procedure that may be adopted is to avoid *heroic* treatment; for though thereby the fat may be diminished, the result may be attained by establishing a worse state of the body.'

* ' Dr. Dalton states that the entire quantity of food· required every twenty-four hours by a man in full health and taking free exercise is of meat 16 oz., bread 19 oz., fat 3½ oz., and water 52 oz.' Jackson, the prize-fighter, in training lived on meat and stale bread only, with of course water.'

† This applies with equal force to animals. The lion and tiger are stronger, fiercer, and more active than the ox or the sheep.

‡ ' If by accident the queen bee dies or is lost, the working bees (which are sexually undeveloped) select two or three eggs, which they hatch in large cells, and then feed the maggots on a stimulating jelly different to that supplied to the other maggots, thus producing a queen bee.'— NEWSHOLME.

13. The following, then, are the objects aimed at in the rules inculcated and the diet laid down in the following pages :   •

1. To improve by exercise the muscular tissue, and by diet to keep the muscles of the body in firm fibre and tone.
2. To maintain the blood in its normal and healthy composition.
3. To regulate the quantity of fluid in the body, by freeing the action of the skin and kidneys.
4. To prevent the deposit of fat, by eliminating from the diet an excess of those articles which create it, but are not otherwise useful in the economy.
5. To allow quite sufficient food, and even many luxuries, so as to satisfy the cravings of nature and the wants of the system, and to do this in a gradual, harmless manner.†

FOOD REQUIRED, AMOUNT OF.

14. 'It may fairly be concluded,' says Dr. Pavy,‡ 'that the requirements as regards food vary with exposure to different conditions. According to the expenditure that is taking place, so in a good scheme of dieting should materials be supplied which are best calculated to yield what is wanted. Under exposure to hard labour and inactivity, and to a high and low temperature, the consumption of material in the system differs, and the supply of food should be regulated accordingly.'

* Luigi Cornaro, a Venetian gentleman of the seventeenth century, after a wild youth which destroyed his health, restored himself after the age of forty to perfect health by a most rigid diet. He had each day a careful allowance of 12 oz. of food—bread, meat, and yolk of egg—and 14 oz. of light Italian wine. He wrote his book at eighty-three, and lived on his hermit-fare until nearly a hundred years old, and enjoyed excellent health. His wife, who we may presume fared in the same manner, lived nearly as long.

† A friend of the author's proposes fitting up a sanatorium for treating sufferers from corpulency, dieting in the manner recommended in this work.

‡ 'Food and Dietetics,' by F. W. Pavy, M.D.   J. and A. Churchill, London.

15. 'The laws of nature,' he continues, 'are such as to conduce to an adaptation of the supply of food to the demand. We are all conversant with the fact that exercise and exposure to cold—conditions which increase the demand for food—sharpen the appetite, and thus lead to a larger quantity of material being consumed, while conversely a state of inactivity and a warm climate tell in an opposite manner and reduce the inclination for food. A badly-fed labourer is capable of performing but a slight day's work, and a starving man falls an easy victim to the effects of exposure to cold.

16. 'Practically it is found that hard work is best performed under a liberal supply of nitrogen-containing food (*i.e.*, meat). The reason is that it leads to a better nourished condition of the muscles and the body generally. Under the use of animal food which is characterized by its richness in *flesh*-forming matter, the muscles are observed to be firmer and richer in solid constituents than under subsistence on food of a vegetable nature;' and it is also obvious that under animal food there is not that danger to corpulency which obtains under a vegetable diet, *unless from that vegetable diet* be eliminated those particular' articles that contain too large a percentage of sugar and starch. To sum up, science teaches us that a liberal supply of meat is necessary to maintain muscles in a good condition for work, as exercise is to make them firm and red, and the result of experience tends to confirm it.

17. None the less is it necessary to limit a meat diet and *dilute* it with a proper admixture of vegetable and other material, and only where it becomes a question of reducing corpulency does it matter what vegetable is taken for this purpose or relatively the amount. In these pages those vegetables only are given the percentage in which of the fattening principles is not in excess of what the system can bear. A large and varied choice still remains to satisfy the epicure and the gourmet.

18. In any dietary for the reduction of corpulence, it is necessary to bear this in mind, that there is a limit beyond which it is not safe to go, and under no circumstances should the loss of weight exceed one pound per week. If this should be so, the diet is not properly adjusted with regard to its constituents, but is simply slow starvation, a result not aimed at in these pages, or desirable.

19. Let us proceed to consider the amount of mixed food necessary to sustain life, health, and strength under different circumstances. This can be calculated to a nicety, but the amount taken must have some relation to the amount of muscular work that the individual is called upon to exercise, the season, and a few other surroundings.*

20. It will be asked, What is the ordinary amount of mixed food an ordinary-sized person should take? An average healthy male adult of medium weight and height, and performing a moderate amount of work, requires

$4\frac{1}{2}$ oz. of nitrogenous food,†
3 oz. of fats (hydro-carbons),
$14\frac{1}{2}$ oz. of carbo-hydrates.‡
1 oz. of salts.

This diet is equivalent to a little over 46 oz. of *moist* solid food.

21. The above ingredients would be contained in $\frac{3}{4}$ of a lb. of meat and a little less than 2 lb. of bread, or in 17 hens' eggs of ordinary size, supposing eggs only were eaten, which it is needless to say would be impossible.

---

* Prison diet where hard labour is done consists of 184 oz. of solid dry food per week. This 184 oz. is made up of meat, bread, cocoa, oatmeal, milk, treacle, barley-meal, salt, cheese, flour, suet, carrots, onions, potatoes. This would mean about 52 oz. of *moist* food per day; for of course the water is not reckoned when the amount of food is chemically considered. Banting's dietary is about 10 oz. a day of *dry food*, which is bare subsistence diet, but it contains a large proportion of muscle-forming ingredients too large for continued safety.

† This embraces meat of all sorts—eggs, milk, and certain constituents of vegetables.

‡ Carbo-hydrates mean bread, sugar, and all vegetables containing starch and sugar.

22. Dr. Lyon Playfair has estimated the quantity of diet required under varying conditions of work as under :

	NITROGENOUS.	CARBONACEOUS.
Subsistence only	2·0 oz.	13·3 oz.
Quietude	2·5 ,,	14·5 ,,
Moderate exercise	4·2 ,,	23·2 ,,
Active work	5·5 ,,	26·3 ,,
Hard work	6·5 ,,	26·3 ,,

23. It will thus be seen that subsistence diet would be represented by about 28 oz. of ordinary moist food per day, and hard work diet by about 60 oz. of the same, and here we will take the amount of food allowed by those who, like Banting, form a dietary for the reduction of corpulency.

24. It is admitted that the human body decreases in fat if the daily food consists of the three great groups of food in the following proportions :

Albuminous food about $4\frac{1}{2}$ oz.,

Fatty food $1\frac{2}{5}$ oz.,

Starchy food (carbo-hydrates) $5\frac{1}{3}$ oz.

This means, under ordinary circumstances, about 22 oz. of moist food daily, and this is not sufficient in amount for safety.

25. Banting, Ebstein, and Oertel, the three men whose systems have been largely adopted for reducing obesity, advocate the different foods in the following proportions :

	ALBUMINOUS.	FAT.	CARBO-HYDRATES.
Banting	6	$\frac{1}{3}$	$2\frac{3}{4}$
Ebstein	$3\frac{1}{2}$	3	$1\frac{1}{4}$
Oertel	$5\frac{1}{2}$ to 6	$1\frac{1}{4}$	$2\frac{1}{2}$ to $3\frac{1}{2}$

26. 'In taking appetite as a guide in regulating the supply of food,' says Dr. Pavy, 'it must not be confounded with a desire to gratify the palate. When food is not eaten too quickly and the diet is simple, a timely warning is afforded by the sense of satisfaction experienced as soon as enough has been taken, and not only does a disinclination arise, but the stomach even refuses it if this amount be far exceeded. With a variety of food, however, and especially food of an agreeable character to the taste, the case is different. Satiated with one article, the stomach is still ready for another, and thus, for the gratification of taste, and not to satisfy appetite, men are tempted to consume far more than is required, and also, it must be said, far more than is advantageous to health.'

27. Hospital diet furnishes a fair estimate of what is necessary under ordinary circumstances, and, taking Guy's as an example, it is found that the daily allowance is 29½ oz. of solid food, apart from the liquids supplied. This amount would represent 16¾ oz. of water-free material. The food actually supplied consists of 4 oz. of cooked meat, 12 oz. of bread, 8 oz. of potatoes, 1 oz. of butter, ¾ oz. of sugar, ¼ oz. of tea, and 3½ oz. of rice-pudding, made of rice, sugar, and milk. There is also a daily allowance of ½ pint of porter and 2½ oz. of milk. This diet is sufficient for the wants of the system under a condition of freedom from labour.

28. Supposing this quantity of food were used for the reduction of fat it would have to be given something in this way, and this will show the difference in the constituents of the food according to the use it is required for:

12 oz. of meat (see page 28),
3 oz. of butter, or fat,
4 oz. of bread, or dry toast,
10 oz. of vegetables,
¼ oz. of tea (sweetened with saccharin),
2 or 3 oz. of milk,

and, instead of porter, a pint of claret, or other light wine (see page 33), daily.

29. In this case the above amount of food would be divided

2

into three meals—breakfast, lunch, and dinner; or breakfast, dinner, and tea, according to the habits of the individual···· and the amount of starchy or farinaceous fcol is here so limited as even in extreme cases to be practically harmless.

30. A diet of this kind would be unendurable if there were no variety; but the endeavour of the author has been to give a large variety, once more reminding the sufferer that food formed on this basis requires a larger amount of exercise to keep the system in health than does the ordinary every-day diet of ordinary people.

31. Taking Mr. Banting's system first. The diet for reducing corpulency advocated by him is now considered to have been wrong, for it threw too much work upon the kidneys and starved the system elsewhere. Banting and his ideas have passed into well-merited oblivion.

32. The Ebstein formula, which has many adherents in Germany, consists in very much restricting the food eaten, giving a large portion of fat, and curtailing those articles of diet that contain sugar and starch, his theory being that fat assists in the body, by its transformation into heat, in eliminating other materials. It is not believed now that fat creates fat. It is believed that by its combustion in the economy it keeps up heat, and at the same time oxidizes waste; in short, it acts much in the same way as, in a railway-engine, the fuel acts which is used in stoking.

33. The system of diet advocated by Oertel not only provides for the gradual decrease of fat, but also prevents its reaccummulation, and at the same time restores tone to the heart, muscular and nervous systems, which excessive corpulency much impairs.

34. A very moderate curtailment on this system will be sufficient to reduce a corpulent person to respectable dimensions, if at the same time active riding or walking exercise be taken. Recreation, by improving the *quality* of the blood, as well as by circulating it more rapidly, assists the oxidation— *i.e.*, the consumption of tissue, especially of fatty tissue—and

that it does this may be shown by the fact that exercise absolutely increases the size of muscles, while it is at the same time reducing weight.

35. There is really no difficulty in reducing corpulence. The difficulty is to get the sufferer—generally easy-going and indolent—to carry out the necessary system, which is more or less irksome by the restraint it puts upon appetite, especially in those who perhaps may almost be said to 'live to eat' instead of 'eating to live.' It is also needless to say that any system, however good, for reducing redundancy of fat would be of no avail if the patient persists in eating and drinking between meals.*

36. Oertel draws two distinctions of obesity, namely, the slight form in which the organs of circulation are unaffected ahd where exercise is possible, and the more serious form, in which fat is deposited in the muscle of the heart, which is thereby dangerously weakened, and, as a corollary, the healthy action of *all* the other organs in the body is interfered with.

37. All who have made the treatment of obesity a study reduce considerably the amount of farinaceous food and sugar, the great offenders in contributing to the accumulation of fat; but all do not allow the same amount of aliment, and some of the dietaries—Banting's, for instance—would be simp'y slow starvation to a man who had to do an ordinary amount of hard mental or bodily work.

38. 'From 22 to 26 oz. of solids and about 35 oz. of liquids per day constituted Mr. Banting's allowance. If we allow for water chemically combined with the food, the daily amount of solids may be set down at from 11 to 13 oz.' Now, this is far from a generous allowance, even if it were of the most fattening materials, and no wonder he got thin on it. The diet tables of prisons, of London needlewomen, and that of the cotton operative during the Lancashire cotton famine,

---

* See 'Aids to Long Life,' by the same author. Chatto and Windus, London, pp. 283.

averaged, of nitrogenous matter (meat), 2·30 oz. ; of fat, ¾ oz. ;
and of starches and sugars, 11½ oz.

39. Moleschott gives 23 oz. of *dry* food as about
the average a healthy man during ordinary work ought
to eat ; this would be equal to nearly twice the amount of
moist food. Assuming that this large amount of food were
deprived of the great fat producers—sugar and starch—it is
not too much to say that, with a fair amount of work, there
would be no great fear of a person becoming obese on it.
A very big man might find it a proper diet.

40. The following figures show what should be the relative
height and weight of a person of adult age in good health :

EXACT STATURE.		MEAN WEIGHT.		
ft.	in.	st.	lb.	lb.
5	1	8	8 or	120
5	2	9	0 „	126
5	3	9	7 „	133
5	4	9	13 „	139
5	5	10	2 „	142
5	6	10	5 „	145
5	7	10	8 „	148
5	8	11	1 „	155
5	9	11	8 „	162
5	10	12	1 „	169
5	11	12	6 „	174
6	0	12	10 „	178

41. It reads thus : a man in his clothes, of 5 feet 8 inches,
should weigh 11 stone 1 lb. ; he may exceed this by 7
per cent., and so attain 11 stone 12 lb. without affecting
his vital capacity ; beyond this amount his respiration becomes diminished.*

---

* Among the Asiatics there is a sect of Brahmins who pride themselves
on their extreme corpulency. Their diet consists of farinaceous vegetables,
milk, sugar, sweetmeats, and ghee. They look upon corpulency as a sign
of opulence ; and many arrive at a great degree of obesity without tasting
anything that has ever lived.

FOOD: ITS USES AND ULTIMATE ELIMINATION.

42. Popularly, we speak of the products we eat to supply the wants of the system as 'food' and 'drink,' and the ordinary individual does not care to analyze further the meaning of the words; but, to the chemist, they have a far broader meaning, and in his hands their different properties and uses in the system are worked out, and in this way, by the light of his knowledge, we are able to show what effects on the animal economy certain foods produce, either for good or harm.

43. The chemist broadly divides foods into two classes. These are known as

THE NITROGENOUS
and the
NON-NITROGENOUS.

The nitrogenous class of foods are those which form the essential basis of structures possessing active or living properties. and the non-nitrogenous principles may be looked upon as supplying the source of power—in other words, if man were looked upon as a steam-engine, the nitrogenous food would form the iron, brass and works of the engine, and the non-nitrogenous would be the coal or any other fuel used in generating power.

44. Now what constitutes nitrogenous food? The answer is: meat of all kinds, gelatine, eggs, milk, and certain constituents of vegetables, such as gluten, vegetable fibrine and caseine.

45. As life consists in the constant renovation and decay of living tissue, and as living tissue, *i.e.*, the body, is made up of nitrogenous matter, it is therefore absolutely necessary for all the operations of life, and is the instrument of living action, and out of it are formed bone, muscle, nerves, etc.

46. Non-nitrogenous* food consists of fats, starch—the basis of bread and all farinaceous foods—sugar, alcohol, and certain vegetable matters. These principles are found either naturally or are produced by chemical action. These constituents are used in the animal economy to keep up the heat of the body, generate power, and when not consumed or eliminated from the system, to be stored up as *fat*.

47. Again, taking a railway-engine as an illustration, it is plain that if rapid speed is required, and a great weight has to be drawn, a greater amount of fuel must be consumed. So, in like manner, if hard work has to be done for many hours a day, a greater amount of food must be taken, and this is consumed in the human body in renovating the tissues and generating the force and heat according to the nature of the food used and its amount.

48. That this is the case may be instanced by this fact amongst many. In making the railway from Paris to Rouen, it was found that two English were equal to three French navvies. An examination of the cause disclosed the fact that the former were fed on large quantities of *meat*,† while the latter ate chiefly *soup* and *lentils*. The diet of the Frenchmen was altered to the English standard, with the result that the inequality soon disappeared.

49. Occupation, season and climate greatly influence the amount and kind of food necessary. The inhabitants of cold climates require a large amount of fat ‡ and in the spring, as vital processes are more active, more food is required. On

---

* Chemists further divide this into hydro-carbons, or fats, and carbo-hydrates, such as starch, sugar, etc., fats being principally heat-producers, and sugar and starch power-producers. This of course is a rough analysis.

† 'Nitrogenous food in this way forming the instrument of living action is incessantly being disintegrated. Becoming thereby effete and useless, a fresh supply is needed to replace that which has fulfilled its office. The primary object of nitrogenous alimentation may therefore be said to be the development and renovation of the living tissues.'—DR. PAVY.

‡ An Esquimaux will eat 10 lb. or 12 lb. of blubber daily; and their children will make wry faces at sugar, but eat blubber with delight.

the other hand, muscular work demands a larger supply of nitrogenous food, *i.e.*, meat.*

50. It will thus be seen that as in the human body certain foods produce muscle, energy, force, and power, others, by their chemical decomposition, furnish material for the production of heat. It is plain that if more is used than is consumed in these ordinary operations of life, or is excreted by the bowels, kidneys, and lungs, it must remain in the system contaminating the blood—as with gout-poison—or equally out of place, and equally destructive to comfort, as fat.

51. On the equable assimilation and excretion of these different classes of food depends the health and comfort of the individual; but from faulty diet, heredity or mode of life in many persons, the balance is not equally held, and the waste that should be excreted, or consumed by exercise or work, becomes stored as fat.†

52. ' The rich may go and have their flues swept out at such places as Carlsbad, Marienbad, Kissingen, and Ems, where the meagre diet and quantity of water drunk (because it is the correct thing to do and the fashion, and because example has such influence for good or evil in this world) will soon reduce their superabundance of unhealthy tissue to limits compatible with health and enjoyment. But the middle-class victim of obesity and gout must be content, if he would enjoy life and live to an advanced age, to take the advice of Socrates, where he says, " Beware of those foods that tempt you to eat when you are not hungry, and of those liquors that tempt you to drink when you are not thirsty," and make dietetics his study to the extent of learning what foods he should indulge in, and what quantity he can take without ultimate discomfort.'

---

* The trappers of the American prairies can live, and do live, for weeks on *meat* and *tea* only.

† Lord Byron undervalued David Hume, denying his claim to genius on account of his bulk, and calling him from the Heroic Epistle ' the fattest hog in Epicurus's sty.' Another of this extraordinary man's allegations was that ' fat is an oily dropsy.' To stave off its visitations he frequently chewed tobacco in lieu of dinner, alleging that it absorbed the gastric juices and prevented hunger.—' Rejected Addresses,' by James and Horace Smith.

USES OF FAT IN THE BODY.

53. Though excess of fat is an evil and an encumbrance, it must not be forgotten that a twentieth part of the male body should be of this substance, and a sixteenth part of that of the female. This may be slightly increased without interfering with the breathing capacity or the comfort of the individual.

54. A moderate amount of fat is one of the signs of health, and is certainly an adjunct to beauty of face and form, and its uses in the animal economy are many and various. In the first place, it serves the merely mechanical purpose of a light, soft, and elastic packing material, which, being deposited between and around the different organs, affords them support and protection from the injurious effects of pressure. Further, being a bad conductor of heat, the fat beneath the skin serves to some extent as a means of retaining the warmth of the body.

55. But the most important use of fat is seen in what occurs during the process of nutrition ; for when more fat-forming material is taken into the system than is absolutely required for the maintenance of the body, it is stored up and laid by to become available for use when the expenditure exceeds the immediate supply.

56. When the direct supply of nourishment is cut off, by withholding it, or by the interruption of the process of digestion, nature has recourse to that which has been laid up in reserve in the form of fat. As everyone knows, in the wasting of the body which ensues as the result of starvation, fat is the first part consumed.* But it has been found by

---

* On the 14th of December, 1810, a pig was buried in its sty by the fall of part of the chalk cliff under Dover Castle. On the 23rd of May, 160 days afterwards, Mr. Mantell, the contractor, was told by some workmen employed in removing the fallen chalk that they heard the whining of the pig, and although he had great doubt of the fact, he urged them to proceed with clearing away the chalk from the sty, and was soon afterwards surprised to see the pig extricated from its confinement alive. At the time of the accident the pig was in a fat condition, and supposed to have weighed about 160 lb. When extricated it presented an extremely emaciated appearance, and weighed no more than 40 lb.

experiment that life cannot be sustained on fat alone. A duck fed on fat only died of starvation at the end of three weeks. Butter, it is said, exuded from all parts of its body, and the feathers seemed as if they had been soaked in melted butter. Similarly animals fed on *fat* and *arrowroot* mixed will die of starvation, or on meat alone ; but if *bone* be given with the meat, it is sufficient to support life for any length of time. This is the reason why wild animals in confinement have bone given them with their meat.

57. The following experiments made will show what part the different constituents of food play in the economy. A couple of rats, which had been nearly brought to the verge of death by restriction to starchy matter and fat, were fed with bread and meat for four days, and then with meat alone. A week after commencing the meat their united weight was 9 oz. 1½ dr., and three weeks later 10 oz. 1 dr. Being now placed on a diet of meat with non-nitrogenous food (starch and fat), a notable improvement occurred ; for in three days' time they weighed 11 oz. ; four days later, 14 oz. 2 dr. ; and a week later still, 14 oz. 4 dr.

58. In another experiment, two rats, weighing 12 oz., were placed on an exclusive diet of lean meat and water. They remained healthy in appearance, but *steadily lost weight, and in a month's time* weighed only 8¾ oz. They were now placed on a miscellaneous diet, and in a week's time weighed 12½ oz.

59. In a third experiment two rats, weighing together 12 oz. 7 dr., were kept upon meat diet exclusively. On the thirteenth day *one of the rats died*, the weight of its body being 2 oz. 8 dr., and that of the other 6 oz. 3 dr. The living one was still kept on the same food, and this died ten days later, the weight of its body being then 5 oz. It will thus be seen that meat alone will not sustain animal life for an indefinite period.

60. What the uses of fat are in the food will be found else-where, but we may remark here that it sustains the heat of the

body. The Esquimaux eat fat as we do bread. Dr. Pavy, in his work on 'Dietetics,' says : 'Travellers have dilated on the large amount of food consumed by the inhabitants of cold, as compared with that consumed by those of temperate or hot climates. Accounts are given which almost appear incredible regarding the enormous quantities of food devoured by dwellers in the Arctic regions. Thus Sir John Ross states that an Esquimaux "perhaps eats twenty pounds of flesh and oil daily." Sir W. Parry, as a matter of curiosity, one day tried how much food an Esquimaux lad, scarcely full-grown, would consume if allowed his full tether. The food was weighed, and, besides fluids, he got through in twenty-four hours 8½ lb. of flesh and 1¾ lb. of bread, and "did not consider the quantity extraordinary."'

61. Sir George Simpson, from his travelling experience in Siberia, says : 'In one highly important particular the Zakuti may safely challenge the rest of the world. They are the best eaters on the face of the earth.' Having heard more on this subject than he could bring himself to believe, he resolved to test the matter by the evidence of his own senses. He procured a couple of men who had, he states, a tolerable reputation in that way, and prepared a dinner for them, consisting of 36 lb. avoirdupois of beef and 18 lb. of butter.

62. By the end of an hour they had got through a half of their allowance in Sir George Simpson's presence. Their stomachs at this time projected 'into a brace of kettledrums.' They were then left in charge of deputies, and Sir George was assured, on returning two hours later, that all had been consumed. He remarks that after such surfeits, the gluttons remain for three or four days in a state of stupor, neither eating nor drinking, and meanwhile are rolled about with a view to the promotion of digestion.

63. 'He who is well fed,' remarks Sir John Ross, 'resists cold better than the man who is stinted, while the starvation from cold follows but too soon a starvation in food.' He says further, 'All experience has shown that a large use of

oil and fat meats is the true secret of life in these frozen countries.' Sir John Franklin also states : 'During the whole of our march we experienced that no quantity of clothing could keep us warm while we fasted; but on those occasions on which we were enabled to go to bed with full stomachs, we passed the night in a warm and comfortable manner.'

64. These remarks will show that the use of fat is not so much to make fat as to keep up the heat of the body by its combustion, and as also to act as a store-house of fuel to draw upon, when required for the operations of life.

EVILS OF OVER-EATING, AND TIME FOR MEALS.

65. There is far more harm done by taking too much food than there is by taking too little, and it is only in very exceptional cases that injury results from the latter cause; whereas an enormous amount of discomfort, disorder, and disease, and even curtailment of life, arise from excess in eating and drinking. Where the individual lives plainly and simply, and only obeys the cravings of nature to the extent of satisfying them, there is no need for weights and scales; but how many are there not, who would be far more comfortable and more healthy if they lived upon a measured amount of food and drink ?

66. There are many other evils beyond corpulence that result from excess in eating, and a badly arranged dietary. Among them may be mentioned a deranged digestion, a loaded tongue, an oppressed stomach, vitiated secretions, a gorged liver, plethora and its consequences, a sluggish brain, with horrible dreams during sleep, and depression when awake.

67. Excess in animal food is as bad as excess in vegetable, if it is combined with an indolent mode of life. Whereas excess of vegetable food of certain kinds leads to obesity, excess of animal food, *if not accompanied by exercise,* leads to the accumulation in the system of the materials that form *gout.*

68. As man is designed by nature to consume a mixed diet, we may proceed to consider when and at what intervals food should be taken. The ordinary custom is that three meals should be taken daily, at intervals of five or six hours apart, and this has been found by experience to be best suited to our requirements. This allows a short period of quiescence for the stomach. The size of the meal should be regulated according to the estimate given on page 17, always having regard to the fact that increased muscular work will allow for increased diet.

69. Sex also and age influence the amount of food required. A woman on the average takes a tenth part less than a man; and during growth more food is necessary to minister to the bodily functions, which are then more active, as well as to supply materials for increase. After forty, the diet, if long life is to be enjoyed, should be sparing.

AMOUNT OF FOOD TO BE CONSUMED.†

70. To prevent the accumulation of fat in those of this predisposition, the following system of diet should be pursued:

71. BREAKFAST.—One large cup of tea or coffee, with a little milk and one or two saccharine tabloids,* with 2 or 3 oz. of bread or dry toast, very thinly buttered, or, instead of butter, 3 or 4 oz. of any light meat or fish, contained in breakfast dishes.

72. LUNCH OR DINNER.—An ordinary dish of any soup found in the section devoted to that aliment; 7 or 8 oz. of roast or boiled meat, fish, or any meat dish that may be chosen in the following pages; an ordinary amount of any vegetable given under that head; a small plate of any non-farinaceous pudding, which see; and 5 or 6 oz. of any fruit, if cooked,

---

* These may be procured at a moderate cost, in any of our larger cities, from the leading druggists and dealers in chemists' supplies, and should be used by the corpulent with all things that require sweetening.

† See page 46.

sweetened with saccharine, with six or eight ounces of any light wine, such as claret.*

73. EVENING MEAL.—Tea, coffee, as at breakfast, or 6 to 8 oz. of light wine, diluted or not; dry toast, or a little bread,† with boiled eggs, fish, or any meat dish that may be chosen ; and, as a nightcap, a glass of whisky and water, with a few gluten biscuits, taken just before retiring to rest.

74. This diet, it is needless to say, may be increased if the subject does a large amount of daily work, but in the same ratio of quality of food.   The reader may refer to Dr. Lyon Playfair's table given in page 16.

75. Indeed, the case may be considered a sad one, if even a very liberal allowance of the cookery in these pages may not be taken with impunity and a gradual reduction of weight.

EXERCISE.

76. In the treatment of corpulency this is of the greatest importance.   The muscle of the heart is strengthened by enforced exercise, and the waste of the system burnt off.

77. The nutrition of the muscles is improved by exercise. The blood which they contain is increased, and in consequence of this increased afflux of blood and the more rapid disintegration going on in the muscles, they become harder and larger, and better able to bear fatigue.

78. The action of the skin is increased, and by perspiration the effete matters in the system are got rid of.   The vital capacity of the lungs is increased by exercise.   Digestion becomes more perfect where exercise is indulged in, and the nervous system is improved in nutrition and power.

79. Dr. Parkes, a well-known authority on the amount of

---

* See section on Wine, page 33.

† Several well-known American manufacturers prepare a gluten bread suitable for fat people, as well as other similar specialties—soups, fruits, gluten biscuits, etc.   It may be mentioned that gluten bread or gluten biscuits do not contain any *fattening* principles, as the starch of the farina is carefully eliminated.

exercise desirable, says the average daily work of a man engaged in manual labour in the open air is equivalent to lifting 250 to 350 tons one foot high. This is a moderate amount, 400 tons being a heavy day's work. The amount of muscular work involved in this may be easily known by remembering that a walk of 20 miles on the level road is equivalent to about 353 tons lifted one foot; and that a walk of 10 miles is equivalent to lifting 247 tons one foot high.*

80. 'We may estimate that every healthy man ought to take an amount of exercise represented by 150 tons raised one foot, which is equal to the work done by walking $8\frac{1}{2}$ to 9 miles on a level road. A certain amount of this exercise is taken in performing one's daily work; but apart from this, outdoor exercise should be taken daily equivalent in amount to a walk of 5 or 6 miles. Less than this is not compatible with robust health.'

81. Exercise should be systematic and regular, not taken by fits and starts, and in corpulent people should be increased gradually. The under-clothing should be of flannel, and chill should be guarded against. Lawn-tennis in summer supplies the best form of exercise. Walking uphill strengthens the heart, and the distance and speed should be increased as palpitation subsides.

82. Riding exercise, where practicable, has a stimulating action on the liver and skin, and may be considered the best form of all exercise. Rowing is also an excellent mode of taking exercise. The more muscles that can be brought into play the better.

83. Exercise should not be taken immediately after food, nor should the individual who is anxious to reduce corpulency gratify the desire for a nap. It is well to have recourse to some light mental or bodily employment, such as billiards, chess, etc., to obviate its occurrence; 'but,' as Dr. Pavy remarks, 'with a natural state of things, there ought to be

* Newsholme.

no strong desire to sleep after a meal.' If there be such, it may be concluded that some fault exists. Sleeping after meals may arise from a sluggish state of the liver or kidneys. In this case medical treatment is called for.

84. When any living part is called into frequent and regular exercise, especially if the system is not yet arrived at full maturity, it is observed to become gradually more and more susceptible of action—to increase in size within certain limits, determined by the constitution, and thereby to gain strength, as indicated by an increased power of enduring fatigue and a greater capacity of withstanding the influences of the common causes of disease, to which previously it would have almost immediately succumbed.

85. The explanation of this, as proved by experiment, is that exercise causes an increased action in the nerves and bloodvessels of the part, by which its vitality is augmented and a greater supply of blood and nervous stimulus is sent to it to sustain and repair the greater waste that is taking place, and also to supply additional substance to fit it for the unusual demands made on it. The results of this process are visibly exemplified in men whose habits or profession lead them to constant muscular exertion—in sportsmen, in blacksmiths, dancers, porters, etc., for instance ; and if it is less manifest in *other parts* of the body *beyond the muscles in view*, it is only from other tissues admitting of less expansion and showing their increased power in a different way.

86. Unless exercise in its ordinary sense be taken, neither respiration nor circulation can fully accomplish the purposes they are intended to serve. Life subsists through a series of motions, and all these should be maintained in regular and adequate exercise ; by so doing the food necessary to sustain the system is taken up, and all that is not required is excreted by the different organs that act independently of the will, so that the balance is evenly kept, and none is unnecessarily stored as fat.

STIMULANTS IN CORPULENCY.

87. What part does alcohol, in its various forms, play in the dietary ?　Chemists who have investigated the effect of alcohol on the system have come to the conclusion that it is not a *food*, and does not in any way make flesh or tissue, but in some cases it seems to increase the value of other foods taken. Whether pure alcohol increases or diminishes fat is still a moot point.　Some physiologists believe that it assists in eliminating waste products, while others hold that it has no effect of this kind whatever, and that it passes out of the body unchanged.

88. That it is injurious beyond a certain amount is certain, and that amount is represented by about a sixth of a pint of spirits, about half a pint of sherry, or a pint of claret or other light wine daily.　As alcohol is contained in all wines and fermented drinks, experiments prove that to the other constituents of these beverages we must look for their fattening properties—thus, in wine, to the sugar ; and in ale, stout, and other fermented liquors, to the sugar and starch they contain.

89. That beer and stout are unsuited to fat people there can be no question.　That the English and Germans who drink largely of beer are more inclined to be corpulent than the Scotch or Irish, who drink more of spirits, is a well-known fact. The stronger beers taken to excess in people of a corpulent habit are also apt to lead to the development of gout and biliousness.

90. It becomes a question now to consider the least injurious form in which alcohol may be partaken of by those who require it, or think they do so.　To the ordinary individual, good wine, properly matured, in moderate quantity, is a harmless and exhilarating drink ;* but to those of a corpu-

* Especially does this apply to that period of life when the powers of elimination are declining, and the individual, from the exigencies of age and failing strength, is unable to take the active exercise necessary to consume the waste products of the body.

lent habit of body, certain wines are a slow poison, and it becomes necessary in a work like this to say a few words on so important a subject.

91. The deleterious wines for people constitutionally disposed to stoutness are those which contain sugar, either by arrest of fermentation or by the addition of sugar, and these are, as a rule, the products of hot countries. In France, Germany, or Hungary, etc., where a cooler climate prevails, fermentation occurs with less rapidity, and is allowed to proceed till it comes to a spontaneous termination.

92. 'Here, then, the transformation of saccharine matter is permitted to go on until it is quite, or nearly lost, and in consequence there is produced a drier or less fruity wine, and one which takes less time to mature.' Wines of this class develop a stronger bouquet and a more acid flavour, and they are admitted to be in every way more suited for stout people of sedentary habit. The wines of the Rhine and the Moselle are noted for the aroma they possess, and the greater amount of acid they contain, and their freedom from sugar. The same applies to some of the wines of the South of France known as clarets and burgundies.

93. Burgundies suit best those who require an excess of alcoholic strength; but they do not keep well, and sometimes undergo a second fermentation, through mouldiness in the bottle. The grand old vintage wines, such as Chambertin,* Clos Vougeot, etc., are sufficiently alcoholized not to decompose, but their strength and price do not fit them for ordinary consumption. The peculiar odour of wall-flowers is characteristic of the choicer burgundies, but they are necessarily of high price.

94. The choice Rhine wines — moselle, chablis, chateau Yquem—bear carriage badly and are best drunk in their own country; but there are plenty of good claret and light wine brands to which this does not apply. The wines to be

---

* The great Napoleon drank this wine, but he was an exceedingly abstemious man.

avoided are sweet wines, such as port, sherry, champagne, tokay, madeira, etc.

95. The wine-drinker, if he is subject to obesity or gout, places his health and comfort in the hands of his wine merchant. So he should choose one of known probity, whose good name would be too valuable a heritage to lose by palming off adulterated wines and *poisonous new* spirits for the sake of present profit.

96. Always remembering that a pint of light wine daily should not be much exceeded, the corpulent person may with his soup drink a glass of sauterne ;* with his fish a glass of Rhine wine, such as Erbach or Steinberg, ; with his meat or game a glass of burgundy, and with his cigar after, some weak whisky or cognac and water, or tea or coffee sweetened with a tabloid of saccharin.

TEA: ITS USE.

97. Tea is not food, and should not be taken as such. Tea taken three or four hours after dinner is valuable, for this is the time that corresponds with the completion of digestion, when, the food having been conveyed away from the stomach, nothing remains but the acid juices employed in digestion. These acid juices create an uneasy sensation in the stomach, and a call is made for something to relieve this uneasiness. Tea fulfils this object better than stimulants ; more than this, it satisfies some unknown want in the system. This refers to the moderate use and enjoyment of tea, but there is a large class who drink an enormous quantity of this beverage, to the undoubted impairment of their health.

98. Those who take it to excess are found principally among the poor. They become pale and bloodless, much

---

* The wines recommended by the author are kept by numerous American dealers in our larger cities. Be sure to purchase of parties whose wares may be trusted to be what they are represented.

given to faintness, nervousness, and depression of spirits, and
suffer excessively from flatulence and loss of appetite. This
is no doubt partly due to poisons used to colour and adulterate
it. Many women ruin their digestive powers by taking large
quantities of weak tea three or four times a day. One form
of indigestion caused by tea deserves special notice, as it is
commonly observed by medical men : the appetite is unim-
paired, and no particularly unpleasant sensations are felt after
meals, but almost as soon as food is taken it seems to pass out
of the stomach into the bowels, causing flatulent, colicky
pains, speedily followed by diarrhœa. Hence, there is a
constant craving for food and a feeling of sinking and
prostration.

99. In some instances, these symptoms only occur in the
morning ; in others, they follow every meal, and lead to serious
loss of flesh, and also to alarm. It is needless to say the
remedy is to take less tea, and add plenty of milk to it, or
for a time to change to coffee or cocoa. In moderate quantity,
tea exerts a very decidedly stimulant and restorative action
on the nervous system, which is aided by the warmth of the
infusion, and is particularly useful in over-fatigued conditions
of the system, and under these circumstances it is infinitely
preferable to alcoholic drinks. Lord Wolseley considers it is
the best drink for exhausted soldiers after a long march.

100. The harmful effects of tea depend a great deal on the
way it is made. If it is allowed to infuse too long, the tannin
and other injurious ingredients of even the best tea are drawn
out, and the infusion becomes bitter and astringent, and unpleasant to the taste. To make tea properly, the teapot
should be warmed, and the water poured over the tea immediately it boils. Five teaspoonfuls of mixed tea should be
put to each quart of boiling water, and it should draw for
eight minutes. Professional tea-tasters are very particular to
use only water which is freshly boiled.

101. In China tea is sometimes infused in a teacup, and
sometimes in the cup from which it is drunk. In Japan the

tea-leaves are ground to powder, and, after infusion in a tea-cup, the mixture is beaten up until it becomes frothy, and then the whole is swallowed. The Chinese drink their tea in a pure state; the Russians take it with lemon-juice; and the Germans often flavour it with rum, cinnamon, or vanilla. In America we know it is customary to add cream, milk, or sugar, but for corpulent people the Russian mode would be the best.

102. The only way by which ordinary people can detect adulteration of tea is by unfolding the leaf after it has been used. In pure tea this is serrated (like the edge of a saw) almost its whole length, and the veins run out from the tendrils nearly to the border and then turn in, so that a distinction is left between them and the border. The leaf may vary in shape and size. The leaves used to adulterate tea are the sloe, hawthorn, beech, and willow. Tea-leaves, when freshly gathered, are destitute of odour and flavour. The pleasant taste and aroma for which they are so highly valued are developed in the process of drying.

103. Either black or green tea may be prepared at will from the same leaves, gathered at the same time and under the same circumstances. It is the lengthened exposure to the air in the process of drying, accompanied by a slight heating and fermentation, that give the dark colour and the distinguishing flavour to the black teas of the shops. The practice of scenting teas for the foreign markets is common in China, and many teas, especially green teas, are artificially coloured with Prussian blue, indigo, and burnt gypsum. It is said that one day an English gentleman at Shanghai, being in conversation with some Chinese from the green-tea country, asked them what reason they had for dyeing the tea, and whether it would not be better without undergoing that process. They acknowledged that tea was much better when prepared without having such ingredients mixed with it, and that they never drank dyed teas themselves; but that, as foreigners seemed to prefer having poisonous ingredients mixed with their tea to make it look pretty, and as these poisons were cheap, tho

Chinese had no objection to supply them, especially as such teas fetched a higher price.

104. Tea being an article of daily and universal use, the following rules should be observed : 1. Don't buy high-priced or highly-flavoured teas, especially if green, as they owe their flavour to noxious matters. 2. Take a good proportion of milk with the infusion. 3. Let the quantity used at each infusion be very moderate. 4. Make the infusion properly with boiling water, and don't let it draw for more than eight minutes. If these rules be followed tea-drinking has a con-servative action on the different structures of the body, checking any disposition to too rapid a change in them by wear and tear, and thereby preventing waste of tissue.

105. Teas free from too great an amount of tannin suit weak stomachs best. The Indian and Assam teas as a rule contain a large amount of this ingredient. The author can strongly recommend the new digestive tea (so called), which has been latterly imported to some extent. This agrees with people who cannot drink other teas, such as dyspeptics and those of nervous temperament.

COFFEE: ITS USE.

106. ' Coffee,' says Dr. Pavy, ' is said to have been in use in Abyssinia from time immemorial, and in Persia from A.D. 875. It was used in Constantinople about the middle of the sixteenth century, in spite of the violent opposition of the priests, and in 1554 two coffee-houses were opened in that city. It was intro-duced into Europe in the seventeenth century. It was drunk in Venice soon after 1615, and brought into England and France about forty years after.'

107. Like tea, coffee produces an invigorating and stimu-lant effect, without being followed by any depression, and fully justifies the estimation in which it is held. It increases the action of the pulse, and is more heating than tea, while at the same time it arouses the mental faculties and so dis-

· poses to wakefulness.   To make the infusion properly 2 oz. of freshly-ground coffee should be used to each pint of boiling water.

108.   Coffee is especially useful to those who suffer from redundancy of fat, as it has the power of relieving the sensation of hunger and fatigue, and may be used two or three times a day as a beverage.   It has all the advantages of a stimulant without the ill-effects following alcohol in its various forms. It exerts a marked sustaining influence under fatigue and privation, and sustains the strength where a restricted diet is necessary, and this enables arduous exertion to be better borne under the existence of abstinence or a · deficiency of food.

COCOA.

109.   Of this article, as a dietetic for fat people, one need say but little, as it is not suitable.   It is highly nourishing, and contains a large quantity of oil and some starch.   It has, for fat people's use, none of the good qualities of tea or coffee.

WATER AND AERATED DRINKS, ETC.

110.   A supply of water in one shape or another is one of the essential conditions of life.   It is as important as food, and is required for various purposes in the performance of the operations of life.   It forms the liquid element of the secretions, and thereby the medium for dissolving the digested food, and enabling it to pass into the system and the effete products to pass out in solution.

111.   The quantity of water required for drinking purposes is found to bear a relation to climate and to the weight of the individual, being nearly half an ounce for every pound, or one and a half gills for every stone weight.   Thus a man weighing 150 lb. (ordinarily a man of 5 feet 7 inches) would require three pints and three-quarters ; of this about one-third is

taken *in the food*, the remainder, two and a half pints, being required as drink.

112. Where there is a tendency to fat it is not advisable to drink more fluid than is necessary to quench thirst. Soda and potash waters may be taken for this purpose, but no aerated waters that contain *sugar*, such as lemonade, ginger-beer, and their allies. Recipes will be found on pages 123-126 for non-alcoholic drinks, sweetened with saccharin, for such people. During hot weather, even under any circumstances, more fluid is necessary.

113. If a plain and wholesome liquid be drunk, such as tea, coffee, and those recommended under the sections on wine and beverages, the error is not likely to be committed of taking too much. After compensating for what is given off by the skin and the lungs the remainder passes off by the kidneys, and washes away the effete products of the system in a dissolved state. The poison of gout is especially eliminated by the kidneys. Gout is one of the commonest complications of corpulency.

PART II.

DIETETICS AND COOKERY.

114. THE following articles of diet may be partaken of by corpulent people. In the case of meats and fish, all superfluous fat should be removed. (See American Introduction.)

JANUARY.

MEAT.—The lean of beef, mutton, doe-venison.

GAME AND POULTRY.—Hares, rabbits, pheasants, partridges, woodcocks, snipes, fowls, chickens, capons, pullets, grouse, wild fowls, turkeys, tame pigeons.

FISH.—Turbot, soles, flounders, plaice, skate, whitings, cod, haddocks, herrings, oysters, lobsters, crabs, prawns, tench, perch, mussels.

VEGETABLES.—Cabbages, broccoli, savoys, endive, sprouts, Scotch kale, sea-kale, spinach, lettuces, celery, cardoons, salsify, turnips, Jerusalem artichokes, garlic, shallots, mustard and cress, cucumbers, mushrooms.

FRUITS.—Apples, medlars, currants, grapes, walnuts, nuts, filberts, oranges, lemons.*

ESPECIALLY IN SEASON IN JANUARY.—Haddocks, whitings, tench, skate, hares, rabbits.

FEBRUARY.

115. MEAT.—Beef, mutton, venison.

GAME AND POULTRY.—Hares, rabbits, pheasants, partridges,

* As *uncooked* fruits contain sugar they should be taken sparingly.

woodcocks, snipes, pigeons, turkeys, fowls, pullets, capons, chickens, turkey-poults.

Fish.—Flounders, brill, plaice, skate, soles, turbot, codfish, whitings, sturgeon, haddocks, oysters, mussels, cockles, crabs, crayfish, prawns, shrimps, barbels, perch, pike, tench.

Vegetables.—Broccoli, cabbages, Brussels sprouts, savoys, celery, cardoons, lettuces, endive, spinach, sorrel, forced French beans, turnips, and all small salads; tarragon, scorzonera, cucumbers, mushrooms.

Fruits.—Apples, grapes, oranges, pomeloes, shaddocks, almonds, nuts, chestnuts, walnuts, figs, currants, filberts.

Especially in Season in February.—Skate, dace, turkey-poults.

March.

116. Meat.—Beef, mutton, doe-venison.

Game and Poultry.—Fowls, chickens, turkeys, pigeons, rabbits, guinea-fowls, woodcocks, snipe.

Fish.—Turbot, whitings, soles, plaice, flounders, skate, oysters, lobsters, crabs, prawns, cod, crayfish, mackerel, mussels, trout.

Vegetables.—Savoys, cabbages, sprouts, spinach, lettuces, turnips, radishes, Jerusalem artichokes, parsley and other garden herbs, Scotch kale, broccoli, scorzonera, salsify, sea-kale, chives, celery, cress, mustard, sorrel, horse-radish, rhubarb, shallots, cucumbers.

Fruits.—Apples, oranges, forced strawberries.

Especially in Season in March. — Mackerel, mullet, skate, whitings, prawns.

April.

117. Meat.—Beef, mutton, grass-lamb.

Game and Poultry.—Pullets, chickens, leverets, fowls, pigeons, wood-pigeons, rabbits, turkey-poults.

Fish.—Brill, cockles, cod, crabs, dory, flounders, halibut, ling, lobsters, mullet, mackerel, mussels, perch, oysters, pike, plaice, prawns, shrimps, skate, sturgeon, soles, whitings, turbot, trout; shad and its roes.

VEGETABLES.—Asparagus, beans, fennel, endive, broccoli, cucumbers, chervil, lettuces, parsley, rhubarb, turnips, sorrel, sea-kale, radishes, spinach, turnip-tops, small salad, parsnips.

FRUITS.—Apples, oranges, early strawberries, walnuts.

ESPECIALLY IN SEASON IN APRIL.—Prawns, crabs, obsters, grass-lamb, asparagus, cucumbers.

MAY.

118. MEAT.—Beef, mutton, grass-lamb, calf's liver.

GAME AND POULTRY.—Fowls, pigeons, pullets, chickens, wood-pigeons, leverets, rabbits.

FISH.—Cod, crabs, brill, flounders, lobsters, mackerel, perch, prawns, plaice, pike, shrimps, whitings, crayfish, gurnet, dory, haddocks, soles, halibut, turbot, trout, shad, shad-roes.

VEGETABLES.—Cabbage, asparagus, kidney-beans, chervil, turnips, spinach, sorrel, sea-kale, lettuces, rhubarb, corn salad, cucumbers, cauliflowers, radishes, artichokes, salads generally.

FRUITS. — Apples, cherries, currants, strawberries, gooseberries.

ESPECIALLY IN SEASON IN MAY.—Prawns, crabs, lobsters.

JUNE.

119. MEAT.—Beef, mutton, grass-lamb, buck-venison, calf's liver.

GAME AND POULTRY.—Fowls, chickens, pullets, turkey-poults, pigeons, leverets, plovers, rabbits.

FISH.—Turbot, soles, mackerel, carps, pike, crabs, tench, prawns, lobsters, shrimps, mullet, haddocks, trout.

VEGETABLES.—Cauliflowers, spinach, beans, asparagus, artichokes, turnips, lettuces, cucumbers, radishes, cresses, all kinds of salad, sorrel, horse-radish, rhubarb, vegetable-marrows.

FRUITS.—Gooseberries, currants, cherries, strawberries, apricots, peaches, apples, nectarines, grapes, pine-apples.

ESPECIALLY IN SEASON IN JUNE.—Skate, prawns, lobsters, crabs, grass-lamb, vegetable-marrows.

JULY.

120. MEAT.—Beef, mutton, grass-lamb, buck-venison, veal.

GAME AND POULTRY.—Fowls, chickens, pullets, turkey-poults, tame rabbits, wild rabbits, leverets, plovers, wheatears, wild chickens, pigeons, wood-pigeons.

FISH.—Dace, dory, cod, carp, brill, barbel, crabs, crayfish, flounders, haddocks, ling, mackerel, lobsters, mullet, thornback, plaice, pike, soles, tench, gurnet, perch, dabs, prawns, whitings, trout.

VEGETABLES. — Kidney, Windsor, and scarlet beans, asparagus, artichokes, celery, endive, chervil, lettuces, mushrooms, salsify, spinach, sorrel, radishes, turnips, salad, peas.

FRUITS.—Apples, oranges, pine-apples, currants, cherries, damsons, gooseberries, strawberries, raspberries, plums, peaches, nectarines.

AUGUST.

121. MEAT.—Beef, mutton, grass-lamb, venison, veal.

GAME AND POULTRY. — Grouse, pullets, fowls, pigeons, turkey-poults, moor-game, chickens, plovers, turkeys, wild pigeons, rabbits, wheatears, leverets.

FISH.—Turbot, whitings, dace, dabs, tench, thornback, flounders, perch, haddocks, herrings, lobsters, crabs, pike, plaice, barbel, oysters, prawns, gurnet, brill, cod, crayfish, mullet, mackerel, soles, trout.

VEGETABLES.—French, kidney, Lima, and scarlet beans, artichokes, lettuces, cauliflowers, cucumbers, salsify, radishes, salad, mushrooms, shallots, turnips, spinach, leeks, endive, peas, tomatoes.

FRUITS. — Apples, plums, peaches, greengages, damsons, cherries, currants, raspberries, gooseberries, nectarines, filberts.

ESPECIALLY IN SEASON IN AUGUST. — Turbot, mackerel, pike, perch, prawns, dace, crabs, herrings, lobsters, grouse, greengages, filberts, figs.

SEPTEMBER.

122. MEAT.—Beef, veal, mutton, lamb, venison.

GAME AND POULTRY.—Fowls, pullets, chickens, wild duck, partridges, hares, pigeons, rabbits, turkey-poults.

FISH. — Cod, haddocks, flounders, plaice, soles, mullets, lobsters, oysters, prawns, carp, pike, perch, tench, herrings, brill, turbot, crabs, dace, trout.

VEGETABLES.—Cauliflowers, cabbages, turnips, peas, beans, artichokes, mushrooms, lettuces, tomatoes.

FRUITS.—Apples, plums, cherries, peaches, grapes, straw-berries, pines, walnuts, filberts, hazel-nuts, quinces, medlars, currants, damsons.

ESPECIALLY IN SEASON IN SEPTEMBER. — Pike, perch, lobsters, dace, crabs, mussels, hares, moor-game, partridges, grouse.

OCTOBER.

123. MEAT.—Beef, veal, mutton, lamb, venison.

GAME AND POULTRY.—Turkeys, pullets, fowls, chickens, widgeons, larks, woodcocks, grouse, pheasants, pigeons, part-ridges, snipes, hares, rabbits.

FISH.—Oysters, lobsters, crabs, brill, gurnet, dory, smelts, halibut, gudgeon, barbel, perch, carp, tench, herrings, hake, pike, dace, trout.

VEGETABLES.—Turnips, cauliflowers, cabbages, beans, leeks, spinach, endive, celery, scorzonera, cardoon, parsley, salads, garlic, shallots, tomatoes.

FRUITS.—Plums, apples, peaches, medlars, walnuts, filberts, nuts, quinces, damsons, pineapples.

ESPECIALLY IN SEASON IN OCTOBER. — Dace, pike, hake, dory, pheasants, partridges, widgeons, broccoli, truffles, grapes, medlars, tomatoes, hazel-nuts.

NOVEMBER.

124. MEAT.—Beef, mutton, venison.

GAME AND POULTRY.—Hares, rabbits, pheasants, partridges, fowls, pullets, turkeys, widgeons, snipe, woodcocks, larks, pigeons, grouse.

FISH.—Oysters, crabs, lobsters, dory, soles, smelt, gurnet, brill, carp, barbel, halibut, pike, tench, cockles, mussels, turbot, herrings, haddocks, skate, whitings, cod, dace.

VEGETABLES.—Turnips, leeks, shallots, Jerusalem artichokes, cabbages, broccoli, savoys, spinach, beet, cardoons, chervil, endive, lettuces, salsify, scorzonera, Scotch kale, celery, mushrooms, tarragon, parsley, salads.

FRUITS.—Apples, quinces, walnuts, filberts, nuts.

ESPECIALLY IN SEASON IN NOVEMBER.—Pike, tench, plaice, dory, grouse, hares, snipes, woodcocks, chestnuts.

DECEMBER.

125. MEAT.—Beef, veal, mutton, doe-venison.

POULTRY AND GAME.—Hares, rabbits, pheasants, grouse, partridges, woodcocks, snipes, fowls, pullets, chickens, turkeys, widgeons, pea-fowl, larks, capons.

FISH.—Sturgeon, turbot, soles, skate, codfish, haddocks, smelts, dory, gurnet, herrings, sprats, oysters, mussels, cockles, lobsters, shellfish, perch, carp, ling, dace.

VEGETABLES.—Cabbages, broccoli, savoys, Brussels sprouts, Scotch kale, sea-kale, spinach, endive, cardoons, lettuces, skirret, salsify, scorzonera, sorrel, turnips, Jerusalem artichokes, celery, shallots, mushrooms, parsley, horseradish.

FRUITS.—Apples, medlars, figs, filberts, nuts, walnuts, currants.

ESPECIALLY IN SEASON IN DECEMBER. — Haddocks, dace, tench, cod, dory, ling, skate, turbot, capon, pea-fowl.

EXAMPLES OF DIETARY.

FIRST DAY.

Breakfast.

Tea with Saccharin and Cream, or Coffee in the same way.
Bread, stale, 2 oz.
Mutton Kidneys, fried (126), or Broiled Mackerel (136).

Dinner.

Julienne Soup (171).
Gray Mullet, broiled, with Gooseberry Sauce.
Stewed Pigeons and Mushrooms (239), or Roast Rabbit (226).
Asparagus, boiled (399).   French Beans (300).
Claret Jelly (342).

Tea or Supper.*

Tea with Saccharin and Cream, or some light Wine or weak
Spirits-and-water.
Dry Toast or Gluten Bread.*
Oysters (197).   Irish Sandwiches (133).

———————

SECOND DAY.

Breakfast.

Tea with Saccharin and Cream, or Coffee in the same way.
Bread, stale, 2 oz.
Fish Rissoles (150), or Eggs and Mushrooms (161).

* See page 29.   † See page 28.

Dinner.

Pheasant Soup (176).
Boiled Soles (221).
Mutton, neck of, boiled (255).   Pheasant, Salmi of (247).
Mashed Artichokes (323).   Green Peas, boiled (305).
Damsons, Compôte of (360).

Tea or Supper.

Tea with Saccharin and Cream, or some light Wine or
Lemonade.
Dry Toast or Gluten Biscuits.
Kidney Omelette (164).   Potted Pheasant (168).

Third Day.

Breakfast.

Tea or Coffee, with Saccharin and Cream.
Thin Toast, with Indian Devil Mixture (142).
Brawn (148).

Dinner.

Rabbit Soup (187).
Whitings, fried (205).
Mutton, kebobbed (256).   Partridges, broiled (250).
Mashed Turnips (310).   Curried Tomatoes (327).
Stewed Apples (13).

Tea or Supper.

Tea with Saccharin and Cream, or some light Wine or weak
Spirits-and-water.
Savoury Omelette (143).   Chicken à la Marengo (272).

Fourth Day.

Breakfast.

Tea or Coffee, with Saccharin and Cream.
Thin Toast or Gluten Bread.
Mackerel, boiled (135).   Grilled Mushrooms (130).

Dinner.

Oxtail Soup, clear (175).
Crimped Cod (224) with Tartare Sauce (378).
Minced Beef (279).   Guinea-fowl, roasted (235).
Sea-kale, boiled (291).   Cardoons, boiled (316).
Strawberry Jelly (348).

Tea or Supper.

Tea with Saccharin and Cream, or some light Wine or
Lemonade.
Oyster Fritters (223).   Lobster, Sauce Mayonnaise (376).

Fifth Day.

Breakfast.

Tea or Coffee, with Saccharin and Cream.
Stale Bread or Thin Dry Toast.
Sweetbreads, browned (154).   Bloaters (159).

Dinner.

Giblet Soup.
Whitings, aux Fines Herbes (207).
Sweetbreads with Piquante Sauce (241).   Onions with Beef
Steak (263).
Spinach with Cream (318).
Custard (352).

Tea or Supper.

Tea with Saccharin and Cream, or some light Wine or weak
Spirits-and-water.
Cold Game (385). Sheep's Tongues, stewed (149).

SIXTH DAY.

Breakfast.

Tea or Coffee, with Saccharin and Cream.
Gluten Bread or Thin Toast.
Mutton Kidneys, stewed (132). Dried Haddock (162).

Dinner.

Brown Soup (192).
Plaice, filleted (215).
Oxtail stewed with Spinach (226). Perdrix aux Vin (244).
French Beans (300).
Calf's-foot Jelly (338).

Tea or Supper.

Tea with Saccharin and Cream, or some light Wine or Lemonade.
Stale Bread.
Lobster Salad (203). Compôte of Cherries (349).

SEVENTH DAY

Breakfast.

Tea or Coffee, with Saccharin and Cream.
Gluten Bread or Thin Toast.
Kidney Omelette (164). Broiled Trout (170).

4

Dinner.

Mulligatawny Soup.
Smelts, broiled (214), or Sole aux Vin Blanc (219).
Rabbit à la Tartare (267).   Calf's Liver à la Mode (276).
Stewed Lettuce (293).   Eoiled Turnips (297).
Stewed Prunes (340).

Tea or Supper.
Tea with Saccharin and Cream, or some light Wine or weak
Spirits-and-water.
Cold Salmi of Partridges (252).   Eggs and Garlic (160).
Gooseberry Fool.

These *menus* will admit of a large selection of change, and
of a varied mode of life with regard to the meal and the hour
of taking it; but at least five hours should elapse between
each repast, enough being taken to *satisfy the appetite.* but *no
more.*

BREAKFAST DISHES.

MUTTON KIDNEYS, FRIED.

126. Put the kidneys into a frying-pan with an ounce of
butter, and a little pepper sprinkled over them.   When done
on one side, turn for an equal time on the other.   Remove to
a hot dish, add pepper, cayenne, salt, and a little sauce
(Harvey's or any other), and pour the gravy from the pan
over them.   Serve hot, on thin, dry toast.   Time, seven or
eight minutes.

MUTTON CHOPS.

127. Take chops from the best end of the neck, saw off
about four inches from the top and the chinebones.   Cut away
the skin and gristle from the upper end of the bone, which

will give the cutlet a round, plump appearance. Sprinkle
each chop with salt and pepper, dip them separately into
dissolved butter, and broil over a brisk fire. A trimmed
mutton cutlet of five ounces in weight will require about six
minutes to cook.

MUSHROOM TOAST.

128. Stew over a gentle fire a quart of nicely-prepared
mushrooms (just opened ones), first dissolving one ounce of
butter in the stewpan, and seasoning the mushrooms with
white pepper or cayenne, a saltspoonful of mace powdered;
stir them carefully, and toss them in the pan to prevent burning and until the butter is dried and slightly brown, when
add half a pint of stock, the grated rind of half a lemon, and
a little salt, and stew until the mushrooms are tender.

MUSHROOMS (AU BEURRE).

129. Trim the stems, and rub two pints of button mushrooms with flannel dipped in salt. Put them in a stewpan
with three ounces of good butter slightly browned, and stir
them very gently to get the butter well about them. Shake
the pan over a moderate fire, that the mushrooms may not
settle at the bottom. When they have well imbibed the
butter, add a little pounded mace, salt, and cayenne, and
cover closely by the side of the fire to simmer until tender,
when they will be found excellent without any other addition.
Serve them hot on thin dry toast.

MUSHROOMS, GRILLED.

130. Cut the stalks, peel and score lightly the underside of
large mushroom flaps, which should be firm and fresh gathered.
Season them with pepper and salt, and steep them in a marinade
of oil or *melted butter*. If quite sound, they may be laid on a
gridiron over a slow even fire, and grilled on both sides; but
they are best done in the oven if at all bruised.

MUTTON CUTLETS WITH TOMATO PURÉE.

131. Trim cutlets from well-hung mutton, beat them into shape after removing the chinebone, dip them into dissolved butter, brush them with egg, and sprinkle with breadcrumbs. Fry in boiling fat, and turn them frequently during the frying. Put them on blotting-paper before the fire to drain. Have ready a purée of fresh tomatoes made as follows : Pick a pound of ripe tomatoes, break them open, and put them without their seeds into a stewpan, with an onion or a couple of shallots, sweet herbs and spice, if liked, salt and pepper; stir over a slow fire till the tomatoes can be pulped through a hair-sieve; return the pulp to the stewpan to simmer, add an ounce of butter well worked together in a little flour, and stir in two ounces of meat-glaze. Arrange the cutlets in a circle a little overlapping each other, and fill the centre with the purée.

MUTTON KIDNEYS, STEWED (À LA FRANÇAISE).

132. Remove the skins from half a dozen or less fine mutton kidneys and cut them lengthwise into slices a quarter of an inch in thickness. Season each piece rather highly with salt and cayenne, and dip it into some finely powdered sweet herbs, namely, parsley and thyme—two-thirds of the former and one of the latter. Three or four finely minced shallots may be added if liked. Melt a good-sized piece of butter in the frying-pan and put in the kidneys. Let them brown on both sides. When nearly cooked, dredge a little flour quickly over them, add a quarter of a pint of boiling stock or water, a tablespoonful of mushroom ketchup and the strained juice of half a lemon. When the gravy is just upon the point of boiling, lift out the kidneys, put them on a hot dish, and two tablespoonfuls of claret to the same, let it boil for one minute, then pour it over the meat. Garnish with fried sippets. Time, six minutes to fry the kidneys.

Irish Sandwiches.

133. Cut the meat in very thin slices from partridges, grouse or any game that has been roasted, and shred some celery. Lay the meat on delicately thin fresh toast—it should be crisp and not tough—strew celery over, and season well with Tartar sauce. Serve in squares on a napkin.

Lamb Cutlets, Superlative.

134. Take a tablespoonful of each of the following ingredients, all finely minced : parsley, shallots, mushrooms and lean ham. Put these into a stewpan with an ounce of fresh butter, and stir them over the fire for five minutes. Add a quarter of a pint of sauce, a little pepper and salt, half a dessertspoonful of strained lemon-juice, three grates of nutmeg, and the yolks of two eggs. Stir the sauce over the fire till it thickens, but it must not boil. Partially fry the cutlets, when nearly cold dip them into the above preparation and place them upon ice until the sauce is set. Dip the cutlets in egg, fry, and serve them with a purée of spinach or green peas. Time, twenty minutes to fry the cutlets—ten minutes each time.

Mackerel, Boiled.

135. Wash and clean carefully after removing the roes. The mackerel is in its greatest perfection when it has roe. Lay the fish and roes separately into cold water, and to a gallon of water add from three to four ounces of salt, and two tablespoonfuls of white vinegar ; when at boiling-point skim, and simmer only until done. Much depends on the size of the fish. Remove at once when done, or from their great delicacy of skin they will crack if kept in the water. The usual test, when the eyes start and the tail splits, should be attended to. Serve on a napkin with the roe.

Mackerel, Broiled.

136. Large fresh fish should be procured for broiling. Cleanse the fish thoroughly and dry in a cloth, or hang up in

the air.   Open it down the back, rub the inside with a little
salt and cayenne mixed, and smear with clarified butter or
good oil.   Put it into a thickly buttered paper loosely fastened
at each end, and broil over a clear fire, or it may be broiled
without the paper, though the former mode renders the fish
so cooked more delicate, and not so apt to disagree with the
stomach as when exposed to the fire uncovered.

Mackerel, Broiled, and Tarragon Butter.

137. Remove the inside of the fish through the gills and
vent without opening it.   Wash, clean, dry, and make a deep
incision down the back, lay the fish in a little salad-oil, keep
it well basted for three-quarters of an hour, but cut off the
nose or part of the head and tail before it is steeped in the
oil.   Broil over a clear fire, and when done have ready the
following mixture, with which fill up the incision : Work a
little butter, pepper, salt and tarragon leaves, chopped and
steeped in vinegar together.   When ready, serve the mackerel
with some of the butter spread over it on a hot dish.   Time,
from ten to fifteen minutes to broil.

Meat Rissoles, German.

133. Fry two or three ounces of bacon cut into small squares,
then add the following ingredients : To a well-beaten egg stir
a quarter of a pound of well minced cold meat, a quarter of
a pound of soaked bread (squeezed dry), a tablespoonful of
chopped onion, another of parsley, and pepper and salt to taste.
When these ingredients are fried sufficiently, turn them out
into a basin until nearly cold, then add a couple of eggs,
beat all well together, form the mixture into balls about the
size of an egg and fry for six minutes, or drop them into boiling soup stock or water ; pour gravy over them before serving.
Time to make, half an hour ; sufficient for six balls.

Pigs' Kidneys, Broiled.

139. Split the kidneys lengthwise from the rounded part, without separating them entirely. Peel off the skin, and pass a wooden or metal skewer through them to keep them flat. Sprinkle a little pepper, salt and powdered sage over them, oil them slightly. and broil them over a clear fire, the hollow side first, so that the gravy may be kept in when they are turned. Serve on a hot dish, either with or without maître d'hôtel sauce in a tureen. Time to broil the kidneys, four minutes each side, or more according to size.

Pigs' Kidneys, Fried.

140. Peel the kidneys, cut them into slices, dip them in clarified butter, and afterwards into a mixture made of two finely minced shallots, two tablespoonfuls of chopped parsley, a pinch of powdered thyme, and a little pepper and salt. Fry them in an ounce of butter until they are lightly browned, put them into a hot dish, and mix with the butter two tablespoonfuls of thick brown gravy, and two tablespoonfuls of claret. Boil the sauce, pour it over the kidneys and serve hot. If no sauce is at hand, take the kidneys up, mix a teaspoonful of flour smoothly into the butter in the pan, add a wineglassful of boiling water, and a dessertspoonful of mushroom ketchup, a little salt and cayenne and a glassful of claret. Boil the sauce and strain it over the kidneys. Time to fry six minutes.

Hunter's Beef.

141. Take as lean a piece as can be procured of the flank of beef—the thin end is the best. Take out the bones and rub the meat well every day for a fortnight with a mixture made of one pound of salt, one ounce of saltpetre, one ounce of pounded cloves, and one grated nutmeg. At the end of the time roll it as closely and firmly as possible, and bind it securely with skewers and tape. Just cover it with water, and boil or bake it for five or six hours. Do not loosen the tapes, etc., until the meat is quite cold.

INDIAN DEVIL MIXTURE.

142. To a tablespoonful each of vinegar, ketchup, and chutney-paste, add an ounce of dissolved butter, a dessertspoonful of made mustard, salt, and a small cup of good rich gravy. Blend these ingredients thoroughly, and rub them into the meat. Make all hot together slowly. Time, ten minutes to make hot.

SAVOURY OMELETTE.

143. Beat the yolks of four eggs till almost white; beat the whites of four eggs for the same time; beat the two together for five minutes after adding a teacup of milk, some finely-chopped parsley, a dessertspoonful of pickled mushrooms, minced very small, and a teaspoonful of mixed herbs. Add salt and pepper to taste. Beat well. Fry an onion cut in four pieces in some butter, and when boiling take it out before pouring in the omelette mixture. Hold the pan over a clear fire, with a fish-slice fold the omelette when fried a light brown, and serve in a very hot dish.

BAKED EGGS.

144. Well butter some tin patty-pans; break a fresh egg into each; place a lump of butter, some salt, a little cayenne and a little chopped parsley on the top of each. Put in the oven to bake.

FRIED SOLES, OR FRIED SLICES OF COD.

145. Cleanse the fish, and two hours before they are wanted rub them inside and out with salt; wash and rub them very dry, dip them into egg; fry them in boiling lard, dish on a hot napkin, and garnish with crisped parsley.

GRILLED KIPPERED SALMON.

146. Cut some dried salmon into narrow pieces, about two inches wide and four long; broil them over a clear fire, then rub them over with fresh butter seasoned with lemon-juice and cayenne. Serve very hot.

Devilled Hot Meat.

147. Cut some cold meat in slices, and then rub it with the following mixture : a tablespoonful of ketchup, one of vinegar, two of made mustard, one of salt, and two of butter, four tablespoonfuls of cold gravy, one of curry-paste ; mix all as smooth as possible. When rubbed with the mixture put it to the grill before a good fire. Take what is left of the sauce and make it warm, and pour over the grill before sending it to table.

Brawn.

148. Clean and wash a pig's head of six or seven pounds, and put it into a stewpan with two pounds of lean beef ; cover with cold water, and boil until the bones are easily removed, skimming often. Mince the beef and head as fine as possible, but don't let it get cold. Season with fine cloves, a lot of pepper, salt, and cayenne. Stir briskly together, and put into a cake-mould with a heavy weight on the top. Let it stand for six hours. Dip the mould in boiling water when required, and turn the brawn out on a dish. Decorate with green parsley, and serve cold.

Sheeps' Tongues, Stewed.

149. Put the tongues into cold water, bring to a boil till tender enough to remove the skins. Then split them, and lay them in a stewpan with enough good gravy to cover them. Chop some parsley, some mushrooms and onion, finely ; work a lump of butter with it, and season with pepper and salt to taste. Add it to the gravy, and stew till tender ; then lay them in a dish, strain the gravy, pour it very hot over the tongues, and serve.

FISH RISSOLES.

150. Take the remains of cold salmon, or any white fish, free it from bones and skin, pound in a mortar with pepper, salt, cayenne, and a little butter, until it is a smooth paste, but not too dry; make it up into small round balls; wash these over with beaten egg, roll in very fine breadcrumbs, fry a pale gold colour. Serve piled high on cut paper.

BAKED WHITINGS.

151. Clean four or six whitings well, cut off the heads, season well with pepper and salt; butter a pudding-dish at the bottom, lay in the fish, sprinkle them over with more butter and two tablespoonfuls of any light wine, cover with thick buttered-paper, put them in a moderate oven for half an hour; mix together three tablespoonfuls of chablis, two teaspoonfuls of finely-chopped herbs, one dessertspoonful of mushroom ketchup, two tablespoonfuls of gravy, a little cayenne; boil these together for a few minutes, pour over the fish, return them to the oven, and let them do slowly for half an hour more.

KIDNEY BALLS.

152. Chop a veal kidney and some of the fat, some leek or onion, black pepper, and salt to taste; roll it up with an egg into balls, and fry them.

DEVILED PHEASANTS' OR CHICKENS' LEGS.

153. Cut off legs from cold pheasants, score them with a sharp knife, put pepper and mustard and a little salt into the cuts; broil them with a piece of cold butter on each. Serve very hot.

SWEETBREADS, BROWNED.

154. Soak, blanch, and stew the sweetbreads in as much good and nicely flavoured stock as will barely cover them.

When they have simmered about half an hour take them up, and put them into a round saucepan just large enough to hold them, with a good slice of fresh butter, which has been melted and is just beginning to brown. Turn the sweetbreads over and over till they are equally and brightly browned in every part. Keep them hot by the side of the stove. Thicken the stock in which they are boiled with brown thickening. Flavour with mushroom ketchup and lemon-juice, and add a table-spoonful of light wine. Place the sweetbreads on a dish, and pour the sauce over them.

TROUT, FRIED.

155. Empty, clean, and dry the fish thoroughly; cut off the fins and gills, but leave the heads on. Rub them over with flour, and fry them in plenty of hot fat. When they are brown on one side turn them carefully upon the other. Lift them out and drain them on blotting-paper before the fire. Serve on a hot napkin and garnish with parsley.

RED MULLET, BAKED.

156. Wash the mullet and rub it well with lemon-juice; put it in a tin dish, with a large mushroom finely chopped, two shallots chopped, three thin slices of carrot, and four sprigs of parsley chopped, a saltspoonful of salt, the same of white pepper, a quarter of a pint of chablis; bake in a moderate oven for three-quarters of an hour. Baste constantly with dissolved butter; serve with the sauce poured over the mullet. This recipe is written for a large mullet.

BAKED HERRINGS.

157. Take off the heads of six herrings, put them into a deep dish, and season with a saltspoonful of pepper, a tea-spoonful of salt, a quarter of a grain of cayenne, two cloves, four allspice, six peppercorns, a blade of mace, half an inch of bruised ginger, and a teaspoonful of grated horseradish; add a gill of cold water and a gill of good vinegar. Bake in a slow oven for half an hour. Serve cold, with the sauce strained and a teaspoonful of finely chopped chives added.

LING, FRESH.

158. Take one pound of ling, cut it into pieces three-quarters of an inch thick, rub it with pepper and salt, and put it on the gridiron over a clear fire ; in about ten minutes it will be done. Serve it plain, or with lemon or vinegar.

BLOATERS.

159. Open the bloaters down the back and bone them. Lay the fish one on the other (insides together), and broil over a clear fire. When sent to table they are separated, laid on a hot dish, and rubbed over with a little butter ; or, split up, take out the bakebone, trim off the head, tail, and fins, double the fish over, and broil from five to six minutes over a clear fire.

EGGS AND GARLIC.

160. Pound ten cloves of garlic that have been boiled for twenty minutes—the water having been changed during the boiling—with a couple of anchovies ; put them, when well-pounded, into a stewpan, and add two tablespoonfuls of oil, the beaten yolks of two eggs, a tablespoonful of vinegar, some pepper and salt, and mix all together while being treated. Put the mixture on a dish, and serve with sliced hard-boiled eggs. Four eggs will cut slices enough for this dish. Time, ten minutes to dress eggs ; two or three minutes to warm the mixture.

EGGS AND MUSHROOMS.

161. Cut off the ends and stalks from half a pint of mushroom buttons ; put them into a basin of water with a little lemon-juice as they are done. Drain and slice them with some large onions, which fry in butter. If liked, the onions can be omitted and the mushrooms can be stewed whole.

Put the mushrooms when tender on a dish, break some eggs
upon them to cover the surface, and in doing this be careful
not to break the yolks of the eggs. Season with salt and
pepper, sprinkle browned crumbs on the top, and put the dish
in a hot oven till the eggs are set. Serve immediately. Time
to stew mushrooms, from ten to fifteen minutes.

DRIED HADDOCK.

162. Boil it in a frying-pan, with just enough water to
cover it; put it on a drainer to drain, then put it before the
fire with a large piece of butter on it.

HERRINGS, PICKLED.

163. Take half a pound of salt, half a pound of bay-salt,
one grain of saccharin, and an ounce of saltpetre. Pound
all well together until reduced to a fine powder. Procure the
herrings as fresh as possible, cut off the heads and tails, open
them, and lay them for one hour in brine strong enough to
float an egg. Drain, dry the fish with a soft cloth, and put
them in layers into a deep jar, with a little of the powder
between each layer, and a little both at the top and bottom of
the jar. When the jar is full press it down and cover it
closely. The fish will be ready in three months.

KIDNEY OMELETTE.

164. Take two sheep's kidneys and cut into slices; fry them
over a clear fire for three or four minutes in a little butter;
mince them finely, season with pepper and salt, beat up the
yolks of three eggs and the white of one, add the minced
kidney to them. Put an ounce of butter in an omelette-pan,
let it remain on a slow fire until it bubbles, then pour in the
mixture, and stir briskly for a minute or two until the eggs
are set. Fold the edges of the omelette over neatly, and
turn it carefully upon a hot dish. Serve immediately. If too
much cooled it will be tough.

KIDNEYS, MINCED.

165. Chop some ox kidney into pieces the size of a pea, season them rather highly with salt and cayenne, and fry them in one ounce of hot butter for a quarter of an hour, moving them about frequently in the pan that they may be equally cooked. Moisten the mince with two or three table-spoonfuls of good brown gravy.

BROILED MUTTON KIDNEYS.

166. Split them open lengthwise without dividing them, strip off the skin and fat, run a fine skewer through the points and across the back of the kidneys to keep them flat while broiling, season them with pepper or cayenne, lay them over a clear, brisk fire with the cut sides towards it; turn them in from four to five minutes, and in as many more dish and serve them quickly, with or without a cold maître d'hôtel sauce under them. Season them with pepper and fine salt, and brush a very small quantity of oil or clarified butter over them before they are broiled.

DUCK OLIVES.

167. Cut into two joints the legs of a cold duck or chicken, take off the drum-sticks, mix half a teaspoonful of pepper with five or six teaspoonfuls of breadcrumbs, some mixed herbs, a very finely chopped onion, and two teaspoonfuls of chopped parsley. Cut four thin slices of bacon, sprinkle with the crumbs, roll up each joint of fowl in the bacon, tie securely; make hot in a frying-pan or before the fire. Serve on a piece of dry toast thinly buttered.

POTTED PHEASANT.

168. Roast the birds as for the table, but let them be thoroughly done, for if the gravy be left in the meat will not keep half so well. Raise the flesh of the breast, wings, and merrythought quite clear from the bones; take off the skin, mince, and then pound it very smoothly with about one-third of its weight of fresh butter, or something less if the meat should appear of a proper consistence without the full quantity; season it with salt, mace, and cayenne only, and add these in small portions until the meat is rather highly flavoured with both the last. Put in pots, and pour oiled butter over them.

VEAL CAKE.

169. Cut all the brown off some slices of cold veal, and cut two hard-boiled eggs into slices. Get a pretty mould, lay ham, veal, and eggs in layers, and some chopped parsley and a little pepper between each, and when the mould is full get some strong stock, and fill up the shape. Bake half an hour, and when cold turn out.

BROILED TROUT.

170. When the fish is thoroughly cleaned, wipe it dry in a cloth, and tie it round with thread (to preserve its shape). Melt a quarter of a pound of butter with a tablespoonful of salt, and pour it over the trout till quite covered; let it remain in it for five minutes, then take it out and place on a gridiron over a clear fire, and let it cook gradually for fifteen minutes.

SOUPS.

JULIENNE SOUP.

171. Take three carrots, three turnips, the white part of a heart of celery, three onions, and three leeks. Wash and dry the vegetables, and cut them into thin shreds, which should be not more than one inch in length. Place the shreds in a stewpan with two ounces of butter, and stir them over a slow fire until slightly browned. Pour over them three quarts of clear stock, and simmer gently for an hour, or until the vegetables are tender. Carefully remove the scum and grease, and half an hour before the soup is done enough add two pinches of salt and two pinches of pepper. Julienne is seasonable for nine months of the year only.

MULLIGATAWNY SOUP, RABBIT.

172. Cut up two young rabbits into small pieces, and fry them in butter until they are nearly dressed enough, with four onions sliced finely. Place these in a stewpan, pour in a quart of stock, and simmer for an hour. Then take out the rabbit and strain off the onions, replace the rabbit in the stewpan with two more quarts of stock as good as you wish to make it, and when it boils stir in two tablespoonfuls of curry-powder flavoured as you may prefer.

HUNTER'S SOUP.

173. Partially roast a brace of well-kept partridges, or a partridge and a grouse. Put them rather close to a clear fire, and baste them plentifully. As soon as the outside is well browned, take them up, and when nearly cold cut the meat from the bones into nice fillets, and bruise the bones thoroughly. Cut half a pound of lean ham into dice, and fry these in one ounce of butter, with a sliced carrot, an onion, and a little parsley. Add two quarts of strong beef gravy, the bruised bones, and a little salt and cayenne. Simmer gently for two hours, and then strain the soup. Add the slices of meat and a glass of claret, and let it heat once more without boiling.

MULLIGATAWNY SOUP.

174. This is a soup of any kind flavoured with curry-powder. It is highly stimulating, gives tone and vigour to the digestive organs, and is frequently acceptable in very hot or cold climates. Nevertheless we do not recommend its frequent use, though it may occasionally be resorted to on festive occasions.

Ox-Tail Soup, Clear.

175. Cut a fine, fresh ox-tail into pieces about an inch long, and divide the thick part into four. Wash these pieces, and throw them into boiling water for a quarter of an hour, then drain and wipe them with a soft cloth. Put them into a stewpan with two carrots, an onion stuck with three cloves, a sprig of parsley, a small piece of thyme, two or three sticks of celery, ha'f a blade of mace, a teaspoonful of salt, six or eight peppercorns, and a quart of water or clear stock. Boil, remove the scum carefully as it rises, then draw the saucepan to the side of the fire, and simmer very gently until the meat is tender. Lift out the pieces of ox-tail, and strain the soup.

Pheasant Soup.

176. Flour a well-hung pheasant rather thickly, put it down to a brisk fire, and roast it for a quarter of an hour, basting it plentifully all the time. Take it from the fire and let it get nearly cold, then take off the flesh from the breast and the upper part of the wings, skin it and put it aside. Cut up the rest of the bird, and bruise the bones. Scrape a small carrot finely, put it into a stewpan with an ounce of butter, an ounce of the lean of an unboiled ham finely minced, a small sprig of thyme, a bay-leaf, a handful of parsley, half a blade of mace, three or four cloves, half a dozen peppercorns, a shallot, and three or four of the outer sticks of a head of celery. Stir these ingredients over a gentle fire until they are brightly browned, put in the flesh and the bruised body of the bird, pour over them a quart of veal or beef stock, and after boiling stew gently for half an hour, and be careful to remove the scum as it rises. Strain the soup, and rub the meat through a sieve. Mix the purée with the soup, add to it a small pinch of cayenne, a little salt, a glassful of chablis, and the fillets of the pheasant cut into thick slices. Stir over the fire until it is quite hot, and serve. Time, an hour and a half or more.

5

Pheasant Soup—Another Way.

177. Roast a well-hung pheasant according to the directions given in the last recipe. Take off the flesh from the breast and wings; skin it and lay it aside. Divide the rest of the bird into joints, put it into a stewpan, pour over it a quart of unseasoned beef stock, let it boil and simmer gently for three hours. Strain the soup and carefully rub the meat through a sieve. Pound the flesh of the breast in a mortar until it is reduced to a smooth paste; mix with it an ounce of fresh butter, a large teaspoonful of salt, a blade of mace powdered, and a pinch of cayenne. Mix this paste with the soup, stir it over the fire until it is quite hot, skim carefully, and serve. Time, nearly five hours.

Hare Soup.

178. Take the remains of a hare which has been roasted the day before, add to it a few bits of parsley, a stick of celery, a bunch of sweet herbs, also about a quart of water or weak stock. Simmer gently until the meat is nearly off the bones; strain it, pick the meat off the bones; rub this well through a hair or fine wire sieve; add it to the soup with pepper, salt, and half a glass of port wine.

Fowl or Chicken Soup.

179. Have ready about two quarts of stock from veal bones, put to it some vegetables—carrot, onion, celery, parsley and sweet herbs. Take a fowl or chicken, cut it into four pieces, put it into the stock, boil or simmer gently till tender. Strain the soup; add to it the beaten yolks of two eggs, pepper and salt to taste. Cut some of the best meat off the fowl in neat pieces or joints, and add them also. Warm well and serve. A chicken if young will not require to stew so long as the vegetables.

GAME SOUP.

180. Take any game too old for roasting—a couple of partridges or three moor-fowl ; stew them well and slowly in about three pints of stock.   When tender, take them out, cut off some of the best pieces, return the rest to the soup ; add pepper, salt, and a little ketchup.   Let this simmer gently while you prepare the pieces you have cut off.   Take these pieces, trim them neatly, season well, shake a little flour over and fry a nice brown, but don't let them be greasy.   Strain the soup through a sieve, rubbing as much of the meat through from the game as you can ; return the soup to the pan, put in the fried pieces of game ; make it very hot, and serve.

GREEN PEA SOUP.

181. Take a quart of green peas, a stick of celery, an onion and turnip, all cut, some mint, and about a quart of stock. Stew till tender, then pass through a fine sieve.   If too thick, add some more stock ; add salt, pepper, and a little saccharin ; have ready about a pint of young green peas, parboiled in water with a little mint, add them to the soup, and let the whole simmer until tender, then serve.

OYSTER SOUP.

182. Allow about three dozen to a quart of soup.   Open them carefully, keep and strain the liquor from them, beard the oysters, and put the strained liquor over them.   Take a quart of the palest veal stock and simmer the beards in it for twenty minutes ; strain, adding a little more stock if required. Put the oysters over the fire in their own liquor to plump them, but do not let them boil.   Put the soup over the fire, add mace and cayenne, pour the liquor from the oysters to it, put the oysters into the tureen, and pour the soup over them and serve.

Fish Soup.

183. Take any white or fresh-water fish, cut off some of the best pieces, boil the bones and the other parts of the fish in a quart of water for an hour ; let it get cold, or nearly so ; slice a small onion, then put one ounce of butter into a saucepan, put in the onion, let it brown ; then lay in the pieces of fish which were kept, add pepper, salt, a glass of chablis, a tomato sliced and cored ; pour on to it the liquid from the fish bones, straining it to prevent the bits going in ; add a little chopped parsley ; simmer a quarter of an hour. Serve with fried toast.

Tomato Soup.

184. Put two ounces of butter into a saucepan, slice six tomatoes, two carrots, one onion, four ounces of veal, and one of ham, add to the butter, let it steam a quarter of an hour, then add a good quart of stock (made from bones or beef), pepper, salt and a bunch of sweet herbs. Simmer half an hour, take out the herbs, and pulp the rest though a sieve.

Kidney Soup.

185. Add to the liquor of a boiled leg of mutton a bullock's kidney, put it over the fire, and when half done take out the kidney, and cut it into pieces the size of dice. Add three sticks of celery, three or four turnips, and the same of carrots, all cut small, and a bunch of sweet herbs tied together ; season to your taste with pepper and salt. Let it boil slowly for five or six hours, adding a spoonful of mushroom ketchup. When done, take out the herbs, and serve the vegetables in he soup.

White Soup.

186. Chop up any remains you may have of cold veal, chicken, game or rabbit roasted dry. Grate them, beat them in a mortar, and rub them through a sieve. Then add to the

panada a quart of stock, put it into a stewpan, and pay great
attention to skimming it.

RABBIT SOUP.

187. Skin and empty a fine rabbit, and lay the liver aside.
Cut it into joints, and fry them lightly; put them in the
stewpan with the liver and three pints of good stock made
from bones; let them simmer as gently as possible for an
hour, or until the rabbit is done enough, carefully removing the scum as it rises.   Take out the rabbit, cut off the
best of the meat, lay it in a covered dish, and put it in a
cool place.   Bruise the bones and put them back into the
stock, and with them two onions, a shallot, a carrot, a small
bunch of parsley, a pinch of thyme, three or four outer sticks
of celery, and a little salt and cayenne.   Simmer the broth
two hours longer.   Take out the liver, rub it till smooth with
the back of a wooden spoon, moisten with a little of the
liquor, and return it to the soup.   Just before sending to
table add half a glassful of claret and a teaspoonful of mushroom ketchup.   Cut the pieces of meat into dice, let them get
quite hot without boiling, and serve immediately.   Time,
three hours.

VEGETABLE SOUP.

188. Wash, trim, and cut into shreds an inch long a small
cabbage, two large carrots, two turnips, a head of celery, two
leeks, three onions, a large endive or a lettuce.   Put them into
an iron pot with half a pound of good butter, one or two grains of
saccharin, a teaspoonful of salt, a teaspoonful of curry-powder,
and let them fry till of a good brown colour ; stir constantly
to prevent burning.   Add two quarts of water and boil
moderately fast for two hours, skimming frequently.   Put
into the tureen a wineglassful of marsala and a teaspoonful
of tarragon vinegar.   Pour in the soup and serve at once.

Cabbage Soup.

189. Scald one large cabbage, cut it up and drain it. Line a stewpan with four or five slices of lean bacon, put in the cabbage, also three carrots and two onions; moisten with a little stock, and simmer very gently till the cabbage is tender; add two quarts of stock, stew softly for half an hour, and carefully skim off every particle of fat. Season with salt and pepper to taste, and serve.

Carrot Soup.

190. Scrape and cut out all specks from two pounds of carrots, wash and wipe them dry, and then reduce them into quarter-inch slices. Put three ounces of butter into a large stewpan, and when it is melted add two pounds of the sliced carrots, and let them stew gently for an hour without browning. Add to them two quarts of stock or gravy-soup, and allow them to simmer till tender—say, for nearly an hour. Press them through a strainer with the soup, and add salt and cayenne if required. Boil the whole gently for five minutes, skim well, and serve as hot as possible.

Brilla Soup.

191. Take four pounds of shin of beef, cut off all the meat from the bone in nice square pieces, and boil the bone in four quarts of water for four hours. Strain the liquor, let it cool, and take off the fat; then put the pieces of meat in the cold liquor; cut small three carrots, two turnips, one head of celery, two onions, add them with a large sprig of thyme, salt and pepper to taste, and simmer till the meat is tender. If not brown enough, colour it with browning.

Brown Soup.

192. Take one pint of finely-flavoured stock; if not clear, beat up the white of an egg to a stiff froth, stir it in with the crushed shell, and beat a little. Boil for five minutes, let it stand to settle, then pass carefully through a napkin or tammy. Put it again on the fire in a clean saucepan ; if not dark enough, stir in a little Liebig or some of Nelson's extract of meat. Boil by themselves a few asparagus-tops or other vegetables ; drain them when tender, and put them to the soup as it is about to be served.

GIBLET SOUP.

193. Clean and prepare the giblets of a duck ; put them into a saucepan with one ounce of butter, and brown very slightly ; add half an onion, four peppercorns, a little salt, and a small bunch of herbs. Pour over the whole nearly one quart of hot water; stew gently until the meat is done to rags. Strain, and when perfectly cold remove every particle of fat. Mix this half-and-half with any other stock. Warm, add a dessert spoonful of chablis and a very little saccharin to a pint of this.

IMPERIAL SOUP.

193*. To a gill of clear, well-flavoured stock mix three beaten eggs, two spoonfuls of cream, salt and pepper. Stir the liquid,. and put in a buttered basin or mould. Cover with greased paper that no water may enter, and steam the custard gently till set. When cool turn it out, cut it into· thin slices, and divide these into small diamonds or squares. Serve in a tureen of clear soup. The above will be a sufficient quantity for two quarts of soup.

LOBSTER SOUP.

194. Pick the meat from a large freshly boiled new lobster, cut it into squares, and set it in a cool place until wanted. Take away the brown fin and bag in the head, and beat the small claws, the fins, and the chin in a mortar. Put them into a stewpan, and with them a small onion, a carrot, a bunch

of sweet herbs, a stick of celery, the toasted crust of a French
roll, a small strip of lemon-rind, a teaspoonful of salt, a pinch
of cayenne, and a quart of unseasoned stock.   Simmer all
gently together for three-quarters of an hour, then press the
soup through a tamis and return it again to the saucepan.
Pound the coral to a smooth paste, and mix a little salt, pepper,
and cayenne with it.   Stir these into the soup, add the pieces,
and when quite hot, without boiling, serve.

FISH.

MULLET, GRAY, BROILED.

195. Scale, clean, and take out the gills and inside.   A fish
of about two pounds would be best for this mode of cooking.
Score the mullet on both sides, lay it on a dish, sprinkle with
salt, and pour three tablespoonfuls of oil over it.   Turn on the
dish, drain, and when to be broiled fold in oiled paper or not ;
the fire should be moderate and even.   The scores should not
be more than a quarter of an inch deep.

MULLET, STEWED.

196. Make a sauce as follows : Put in a stewpan three wine-
glassfuls of stock, slice thinly a small carrot and turnip, also
half a small lemon, add a bay-leaf, a blade of mace, and a
bunch of thyme and parsley.   Lay in the fish and stew gently
over a slow fire.   Strain the gravy, season with salt and pepper,
and serve the fish on a hot dish.

OYSTERS, COLD.

197. Oysters are never so excellent as when they are eaten
uncooked, if only they are quite fresh and newly opened.
Thin brown bread and butter is usually served with them, and
either lemon-juice or vinegar and pepper ; but the true lover
of oysters prefers them with nothing but their own gravy.

Oysters, Omelette of.

198. Mince well a dozen fried oysters. Mix with them half a dozen well-beaten eggs ; season the mixture with a small pinch of salt, a saltspoonful of white pepper, and the eighth of a nutmeg grated, and fry the omelette in the usual way. Or beat half a dozen eggs lightly, and fry them in hot fat until they are delicately set. Put three tablespoonfuls of oyster sauce into the centre, fold the omelette over, and serve on a hot dish.

Oysters, Scalloped (A Simple Method).

199. Open and beard a dozen oysters and wash them in their own liquor. Scrape the deep shells and cleanse them thoroughly. Put an oyster in each one, season it with salt and pepper, and sprinkle breadcrumbs thickly upon it. Put some little pieces of butter on the top, arrange the shells on a dish, and bake in a quick oven or in a Dutch oven before a brisk fire until they are lightly browned. Serve very hot. Time to bake, about a quarter of an hour.

Perch fried with Herbs.

200. Take two moderate-sized perch, wash, empty, and scale them carefully, wipe them dry, and lay them on a dish ; sprinkle a little salt and pepper over them, and pour on them six tablespoonfuls of oil. Let them soak for half an hour, and turn them once during that time. Drain them well, and cover them thinly with finely-grated breadcrumbs, seasoned with pepper and salt, and flavoured with a powdered clove or a little grated nutmeg, a tablespoonful of chopped parsley, and a pinch of powdered thyme. Fry them in boiling fat till the fish are brightly browned. Serve on a hot dish ; garnish with parsley, and send piquant sauce to table in a tureen. Time to fry, ten minutes or more, according to size.

Lobster, Cold.

201. Take off the large claws, and crack the shell lightly, without disfiguring the fish; split open the tail with a sharp knife, and dish the fish on a folded napkin, with the head in an upright position in the centre, and the tail and claws arranged neatly round it; garnish with parsley, salt, cayenne, mustard. Salad-oil and vinegar should be eaten with it.

Lobster dressed with Sauce Piquante.

202. Pick the meat from the body and claws of a freshly-boiled cock lobster, and divide it into neat pieces about half an inch square. Take the yolks of three hard-boiled eggs, pound them well, and mix with them a teaspoonful of raw mustard, half a teaspoonful of salt, and half a grain of cayenne; add very gradually by drops at first, beating well between every addition, one tablespoonful of salad-oil, and afterwards two tablespoonfuls of tarragon vinegar and a dessertspoonful of very hot chilli vinegar; pour it just before serving over the lobster, and garnish with parsley.

Lobster Salad.

203. In making lobster salad be careful that the lobster is sweet and fresh, and that the lettuces are crisp and dry. Unless the latter are perfectly free from moisture, the sauce, instead of blending properly, will be liable to float in oily particles on the top. Take the meat of one or two large lobsters, divide it into neat pieces, and season each piece slightly with pepper, salt and vinegar; place a bed of shred lettuce-hearts at the bottom of the dish, put a layer of lobster on top of it, mixed, if liked, with a few slices of cucumber, cover again with lettuce, and repeat until the materials are exhausted. Decorate the borders with any garnish that may suit the taste.

LOBSTER À LA CRÈME.

204. Pick the meat from a large freshly-boiled cock lobster, mince it finely, and put it into a saucepan with half a teaspoonful of salt, a teaspoonful of white pepper, the eighth of a nutmeg grated, two teaspoonfuls of vinegar, and one of light wine. When quite hot, put with it a little fresh butter lightly rolled in flour, and a quarter of a pint of thick cream. Simmer gently for ten minutes, stirring all the time, and when thoroughly heated put the mixture into the shell of the lobster, place it on a neatly folded napkin, and garnish with parsley.

WHITING, FRIED.

205. These fish are generally cleaned and skinned at the shop; if not, they must be done by the cook. The tails should be fastened into the mouth; brush them over well with egg, beaten up, and fry a nice brown in boiling lard; drain and dry well. They will take from seven to ten minutes to fry.

WHITING AU GRATIN.

206. Take two full-sized whitings; empty, scrape, cleanse, and wipe them, then make deep incisions on each side with a sharp knife, to the depth of a quarter of an inch; butter a shallow dish thickly, sprinkle upon it a little pepper and salt, pour gently over them one or two glassfuls of French light wine, and lay upon the fish two tablespoonfuls of finely-minced mushrooms, mixed with a tablespoonful of parsley; melt an ounce of butter, pour it over the fish, sprinkle two tablespoonfuls of bread-raspings on the top, and bake the whitings in the oven. Send a cut lemon only to table with the fish. If it is preferred that the wine should not be used, three or four tablespoonfuls of pale veal gravy may be substituted for it.

Whiting aux Fines Herbes.

207. Clean and skin the fish, and fasten it with its tail in its mouth. Place it on a dish, season with pepper and salt, and sprinkle over it a teaspoonful of mixed sweet herbs in powder; lay little pieces of butter, here and there, thickly upon it, cover with another dish and bake in a moderately-heated oven till done enough. Turn it once or twice that it may be equally cooked, and serve with the sauce poured over it.

Haddocks, Boiled.

208. Make them very clean, scrape the outside, empty them, and wash well; take out the gills, curl them round like whiting, or lay them flat in warm water with some salt in it. Let the water boil, skim it, then simmer for ten to twelve minutes. Serve very hot.

Haddock, Baked.

209. Wash the fish, scrape off the scales, and in emptying it open it as little as possible; sprinkle a little salt, and squeeze the juice of a large lemon upon it; let it stand two or three hours, turning it over once or twice during the time. Mix the finely-grated rind of half a lemon with two ounces of grated breadcrumbs; add half a saltspoonful of salt, half a saltspoonful of pepper, and a quarter of a small nutmeg. Wipe the fish quite dry, brush it over with beaten egg, and strew the seasoned crumbs upon it; put it on a wire drainer in a dish, pour on it a little clarified butter, and bake it in a moderate oven. Baste it two or three times during the process. Send the gravy from the fish to table with it.

Haddock, Fillets of.

210. Divide the flesh from the bone by running the edge of the knife along the side of the spine, and take off the skin.

Dip the fillets in beaten egg, roll them in fine breadcrumbs, and then fry them in hot lard or dripping. When browned on both sides drain from the fat and serve them on a hot dish. Take a tablespoonful of mushrooms, chopped small, a tablespoonful of finely-minced shallots, and a tablespoonful of chopped parsley. Put these into a stewpan with a quarter of a pint of good brown sauce. Simmer for a quarter of an hour, and serve in a tureen.

RED MULLET, BROILED.

211. Wash it well, dry in a cloth; it is not usually scaled or opened. Put it into buttered writing-paper fastened tight at the ends; broil it over a clear fire for about half an hour, or it can be baked in a moderate oven for the same time. Take off the paper, and serve very hot.

RED MULLET, FRIED.

212. Melt two ounces of butter in a frying-pan, put in one good-sized or two small mullet, squeeze the juice of half a lemon over them, and season with pepper and salt. Let them fry over a gentle fire until they are done enough, turning them over when half done, that they may be equally cooked. Serve on a hot dish.

SMELTS.

213. These fish are very delicate and good, if quite fresh. Draw them at the gills, but don't open them; dry well in a cloth; dip them into beaten egg and fine breadcrumbs. Fry in boiling lard or fat for five to ten minutes.

SMELTS, BROILED.

214. Draw carefully and wipe a couple of large smelts, flour them well, and lay them on a gridiron over a gentle fire. When half done turn them carefully upon the other side. When they are done enough, put them on a hot dish, sprinkle

a little salt upon them, and serve immediately. A cut lemon or a little sauce may be sent to table with them if preferred.

PLAICE, FILLETED.

215. Skin the plaice, lay it flat on the table, and with the point of a sharp knife cut right down the backbone. Insert the knife close to the head, slip it under the flesh, and pass it from head to tail; by this means the fillet may be removed entire and smooth, and the fish is ready to be fried or stewed.

STEWED WHITING.

216. Take off the skin and the heads and tails; lay the fish in a stewpan, and season each one with a quarter of a saltspoonful of salt, one grain of white pepper, a quarter of a saltspoonful of mixed sweet herbs in powder, and for the whole (four or six) the grated rind of half a lemon. Pour in a quarter of a pound of dissolved butter, simmer for ten minutes; add a large wineglassful of marsala and the strained juice of a lemon; simmer five minutes more. Place the fish neatly on a hot dish, and pour the sauce over. Send to table immediately.

BOILED WHITING.

217. Whiting should be large for boiling, and with the skin taken off it is more delicate. Put it into boiling water, and simmer from twelve to eighteen minutes, according to the size; skim well. Drain, and serve on a neatly folded napkin.

BOILED PLAICE.

218. Large plaice is best for boiling. Put it into plenty of hot water, with a tablespoonful of salt and a wineglassful of vinegar; boil up quickly, skim, and then simmer gently for twenty or twenty-five minutes.

SOLE AU VIN BLANC.

219. Put the sole, after it has been trimmed, into a fishpan, and with it some slices of onion, a faggot of sweet herbs, a couple of cloves, some peppercorns and salt. Spread some butter over the sole, and pour in enough French white wine to cover it. Let it boil for ten to twenty minutes, according to size of fish. Keep it covered while it is boiling. When it is done, remove the fish ; keep it hot while making the sauce. Strain the liquor, return it to the pan, and add the yolks of one or two eggs, according to the quantity of liquor ; only do not put too much egg; just enough to thicken the sauce is required. Put in a little chopped parsley, pour the sauce over the fish when thoroughly hot, and serve at once.*

Filleted Soles (Italian).

220. Skin and carefully wash the soles, separate the meat from the bone, and divide each fillet in two pieces. Brush them over with white of egg, sprinkle thinly with breadcrumbs and seasoning, and put them in a baking-dish. Place small pieces of butter over the whole, and bake for half an hour. When they are nearly done, squeeze the juice of a lemon over them, and serve on a dish with Italian sauce poured over.

Boiled Sole.

221. Cleanse and wash the fish carefully, cut off the fins, but do not skin it. Lay it in a fish-kettle with sufficient cold water to cover it, salted with a little salt. Let it gradually come to a boil, and keep it simmering for a few minutes, according to the size of the fish. Dish it on a hot napkin after well draining it, and garnish with parsley and cut lemon. Send lobster sauce to table with it.

---

* Dr. Davy says : 'If we give our attention to classed people—classed as to the kind of food they principally subsist on—we shall find that the fish-eating class are especially strong, healthy, and prolific. In no other class than in that of fishers do we see larger families, handsomer women, or more robust and active men. As an article of nourishment, fish does not possess the satisfying and stimulating properties that belong to the flesh of animals and birds. On account of its being less satisfying than meat, the appetite returns at shorter intervals, and a larger quantity is required to be consumed.'

OYSTERS, BAKED.

222. Mix three tablespoonfuls of finely-grated breadcrumbs with half a saltspoonful of salt, a saltspoonful of white pepper, and a quarter of a nutmeg grated. Open a dozen oysters, dip them in beaten egg, roll them in the seasoned crumbs, put each one in its lower shell, and lay a small piece of butter upon it. Place the oysters in the oven or before the fire for a few minutes, until they are quite hot. Before serving squeeze a little lemon-juice upon them.

OYSTER FRITTERS.

223. Open a dozen oysters, and warm them in their own liquor for a minute ; put them aside. Beat two eggs, and mix with them half a tablespoonful of milk. Add a little salt, a quarter of a saltspoonful of pepper, a quarter of a nutmeg grated, a quarter of a saltspoonful of pounded mace, and a quarter of a teaspoonful of grated lemon-rind. Dip the oysters into this batter, and then into finely-grated breadcrumbs. Fry in hot fat until they are brown and crisp. They may be used for garnishing.

COD, CRIMPED.

224. Make some deep cuts as far as the bones on both sides of a perfectly fresh cod, making the cuts at two inches distance, and cut one or two gashes on the cheeks ; then lay the fish in cold water, with a tablespoonful of vinegar in it, for an hour or two. It may afterwards be boiled or fried. If it is to be boiled, it should be plunged at once into boiling water, and then simmered gently. Crimping renders the flesh firmer, and makes it better both to cook and to serve.

GRAVY.

When meat is roasted, it exudes a thick brown essence known as osmazone; this in most houses is allowed to remain in the vessel or tin dish in which the meat or game has been cooked, and is then thrown away as of no use. The proper way to make gravy is to skim off the fat the meat has been basted with, and then pour either stock or boiling water on the osmazone, adding a little salt, and stirring until it is all dissolved off the vessel or basting dish ; by this means a strong meat-flavoured gravy is obtained that has the characteristics of the meat cooked. Those who prefer a flavoured gravy can add Worcester sauce or port wine, according to taste. All fat should be carefully skimmed off before it is sent *in a tureen* to table, where this process is not carried out, a little Liebig's extract dissolved in boiling water is a substitute ; but this has not the flavour of the particular dish cooked, and is not to be compared to the other process. Good gravy means good cooking, a rare thing in most households.

DINNER DISHES.

Ox-tail, Stewed.

225. Take a fine ox-tail, disjoint it, cut it into pieces about one inch and a half long, and divide the thick parts into quarters. Throw these pieces into boiling water, and let them remain for a quarter of an hour. Take them up, wipe them with a soft cloth, and put them into a stewpan with two quarts of stock or water, a large onion stuck with three cloves, three carrots, a bunch of savoury herbs, and a little salt and pepper. Simmer gently until the meat parts easily from the bones, then put the pieces on a hot dish, reduce the gravy, strain it over them and garnish with toasted sippets. A little lemon-juice is by some persons considered an improvement. Time, three hours and a half to stew the tail.

OX-TAIL STEWED WITH SPINACH.

226. Stew the ox-tail according to the directions given in the last recipe. When the meat is tender, lift it out, strain the gravy, and reduce it to half the quantity. Pour it again over the meat, let it simmer a few minutes, then serve the stew, neatly arranged in a circle on a hot dish with spinach in the centre.

BEEF TRICE.

227. Beat and lard a juicy, tender steak of two pounds, lay it into a close-fitting covered stewpan, with equal quantities of water and vinegar. Add a little vegetable, particularly onion, and stew gently for two hours; but do not allow it to burn or stick to the pan; when cold, cut the meat into strips, smear it with beaten egg, and strew over breadcrumbs well seasoned with pepper, shallot and suet. Fry till it is of a light brown colour, which will be in about ten minutes.

BEEFSTEAK, FRIED.

228. If no gridiron is at hand, put some butter or dripping in a frying-pan and let it boil; then lay in a steak of half an inch thick, and move it continually with the side of a knife or steak-tongs to prevent it from burning. When sufficiently well done on one side, which will be seen by the colour being well spread over the meat, turn it on the other, continuing to move it about with the tongs in a similar manner. If a fork must be used, do not stick it into the juicy part of the meat, but into the fat or edge. When done, serve on a hot dish with a little butter and some mushroom ketchup, tomato or other sauce or gravy, as preferred.

CALF'S HEART, ROASTED.

229. Wash the heart thoroughly in several waters, then leave it to soak for half an hour. Wipe it dry, and fill it

with good veal stuffing; tie a piece of oiled paper around it, and roast it before a good fire for an hour and a half, or more, according to the size. Before serving, take off the paper, and baste it well. Send it to table with plenty of good brown gravy.*

CALF'S SWEETBREADS, STEWED.

230. Put two sweetbreads into a stewpan with some nicely flavoured stock, and let them simmer gently for three-quarters of an hour or more. Take them out and place them on a hot dish. Draw the gravy from the fire for a minute or two, and add to it very gradually the yolk of an egg. Put this over a gentle fire until the sauce thickens, but do not allow it to boil. Just before serving, squeeze into it the juice of a lemon.

MINCED COLLOPS.

231. Mince very well about one pound of raw beef (it must be tender and free from all skin or fat), season with salt and pepper. Put it into a saucepan, and stir with a fork frequently while it heats, to prevent its gathering into lumps; it must be perfectly smooth. Continue to simmer it gently for about a quarter of an hour; if it gets too dry, add a small bit of butter and a tablespoonful of gravy, but if properly cooked it should not require this. Serve in a silver dish with hot water under, as it should be served very hot.

BONED SHOULDER OF MUTTON

232. Bone a lean shoulder of mutton (it should be a small lean one for this purpose), cut off the knuckle-end and any loose bits of skin. Take some fresh or tinned oysters, strain the liquor from them and beard them, rinse them again well in the liquor. Lay your mutton on the table on the skin side, sprinkle it over well with a mixture of salt, pepper and mace; then lay in your oysters, roll, and bind

* See Gravy, page 81.

round with tape. Take two ounces of butter, put it into
a saucepan that will hold the mutton, melt it; then add a few
slices of onion, some peppercorns, and a sprig of parsley; let
these brown a little, put in the mutton, add as much stock as
will be about an inch or rather more deep in the pan, cover
down and let it simmer gently for two or three hours, or till
quite tender; strain the gravy, add to it a little ketchup,
warm well, and pour round the mutton.

INDIAN HASH.

233. Take some thin slices of cold mutton or beef, free
from skin or fat, put them into a stewpan with some good
stock; season with salt, pepper, a little mace, and add a bit of
butter, also a slice of lemon without rind or pips; cover the
pan, and simmer for ten minutes, but on no account let it
boil. Grate the yolks of two hard-boiled eggs, rub smooth in
a small basin, add to them a large teaspoonful of made
mustard; take some of the gravy from the pan, put it to the
above until it is a thin paste; pour this back into the pan,
take out the lemon, let all simmer together for about five
minutes. Serve hot.

MUTTON, CURRIED.

234. Cut one pound of tender cold mutton in small square
pieces. Put two ounces of butter into a stewpan, make t
boiling hot, add two ounces of onion finely minced, brown
this slightly. Then add one ounce of curry-powder and one
saltspoonful of salt, stir over the fire until the curry-powder is
well mixed in, then put in the mutton, and enough stock to
keep it all soft, but it must not be liquid. Let it stew gently
for ten minutes or a quarter of an hour.

GUINEA FOWL, ROASTED.

235. These birds should hang as long as they will before
being cooked; then they should be stuffed like a turkey, and
served with gravy and a sauce piquante. They will take
about an hour to roast.

BOILED FOWLS.

236. Take one quart of boiling water and one quart of cold water; clean and truss your fowl carefully. The legs should be drawn, cutting the skin at the first joint, and then put under the skin into the bodies, while the wings should be cut off short and twisted back; no livers or gizzards should be trussed with boiled fowl. Put them into water, mixed as above, that will be about right heat, but it must entirely cover them; skim well when it comes to the boil, then simmer. For a fowl, an hour; for chickens, about half the time.

BROILED CHICKEN.

237. This should first be boiled for about ten minutes; then allow it to become cold, and split into two, wash over with beaten egg, cover with breadcrumbs, and broil it over a clear fire; it will take half an hour, or if a large chicken, longer. Serve very hot. The legs should be trussed like those of a boiled fowl, and should be made as flat as possible. The inside should be put first on the grid.

PIGEONS.

238. These are good roasted or stewed; if roasted, they should be used fresh and well basted, the heads and necks cut off, and trussed like a duck; pour plenty of water through them before trussing, and wipe dry. Put into the inside of each a little butter and a bit of cayenne. They will take almost twenty-five minutes, but if very young not so long.

STEWED PIGEONS AND MUSHROOMS.

239. Put into a saucepan one ounce of fresh butter, cut up two pigeons into small pieces, let it stew a little, but not brown; add one pint of good gravy, one tablespoonful of mushroom ketchup, salt, pepper and cayenne: stir well until it just boils,

then let it simmer well for three-quarters of an hour; add
one or two dozen small mushrooms, and stew ten minutes
longer ; then add two tablespoonfuls of cream. Serve on a hot
dish, putting the mushrooms round the pigeons.

SWEETBREADS.

240. Have one or two very fresh sweetbreads, trim and
half boil them in veal broth ; leave till nearly cold, then wash
them over well with the yolk of an egg, and put them into
fine, dry breadcrumbs, seasoned with salt and pepper ; shake
them to allow any loose crumbs to drop off, then fry very
gently in butter or lard.

SWEETBREADS WITH SAUCE PIQUANTE.

241. The sweetbread must first be blanched thus : Half
boil it, then throw it into cold water for a minute or two, or it
may be left in the water until nearly cool ; butter it all over
and flour well, put it into a Dutch oven before the fire,
keep turning it and basting well with butter until it is nicely
browned, but not dried. It will take three-quarters of an
hour. Serve with the following sauce : Boil together a tablespoonful of chopped onion, same of parsley and of mushrooms,
in one ounce of butter for five minutes, then add half a pint of
good stock ; add salt and cayenne, and stir in last one tablespoonful of vinegar. Boil a few minutes.

LARKS, BROILED.

242. Pick and clean a dozen larks, cut off their heads and
legs, truss them firmly, rub them over with beaten egg, and
strew breadcrumbs and a small pinch of salt over them. Broil
them over a clear fire, and serve them on thin toasted bread.
Time, ten minutes.

LIVER, FRIED.

243. Take one pound of fresh calf's liver, cut it into neat
slices, a quarter of an inch thick ; cover the bottom of the

frying-pan with some clear dripping, a quarter of an inch in depth; place the pan on the fire, and, when the dripping ceases hissing, put in the liver, and in five minutes turn it. When done enough, dish it and serve very hot. Time, a quarter of an hour.

PERDRIX AU VIN.

244. Roast two partridges; put into a stewpan three table-spoonfuls of rich gravy, a glass of claret, salt, pepper, the juice of a lemon, and a little cayenne. Cut up the birds, keeping them very hot. Make the sauce very hot over the fire, and pour over the partridges.

PHEASANT, BOILED.

245. Pick, draw, and singe the pheasant, and truss it firmly, as if for roasting; cover with buttered paper, wrap it in a floured cloth, plunge it into boiling water, and after it has once boiled up draw it to the side, and let it simmer as gently as possible until it is done enough. The more gently it is simmered the better it will look, and the tenderer it will be. Put it on a hot dish, pour a small quantity of sauce over it, and send the rest to table in a tureen. Time to boil, half an hour from the time of boiling for a small young bird; three-quarters of an hour for a larger one; one hour or more for an old one.

PHEASANT, BROILED.

246. Pick, draw and singe the pheasant, and divide it neatly into joints; fry these in a little fat until they are equally and lightly browned all over, drain them well, season with salt and cayenne, and dip them into egg and bread-crumbs. Broil over a clear fire and serve on a hot dish, with mushroom sauce or piquante sauce as an accompaniment The remains of a cold roast pheasant may be treated in this way. Time to broil, about ten minutes.

PHEASANT, SALMI OF.

247. Roast a well-hung pheasant until it is a little more
than half-dressed, then take it from the fire, and when it is
almost cold cut it into neat joints, and carefully remove the
skin and fat. Put the meat aside until wanted, and place the
bones and trimmings in a saucepan with an ounce of fresh
butter, a sprig of thyme, and a bay-leaf, and stir these ingredients over a slow fire until they are lightly browned, then
pour over them half a pint of good brown sauce and a glassful
of sherry. Let them simmer gently for a quarter of an hour ;
strain the gravy, skim it carefully, add a pinch of cayenne and
the juice of half a lemon, and put it back into the saucepan with
the pieces of game. Let them heat very gradually, and on no
account allow them to boil. Pile them on a hot dish, pour
the hot sauce over them, and garnish with fried sippets. If
there is no brown sauce at hand it may be prepared as follows :
Mince finely a quarter of a pound of the lean of an unboiled
ham, and put it into a saucepan with two ounces of fresh butter,
a shallot, a large scraped carrot, two or three mushrooms (if at
hand), a blade of mace, a small sprig of thyme, a handful of
parsley, two cloves, and half a dozen peppercorns. Stir these
over a slow fire until they are brightly browned, then dredge
a tablespoonful of flour over them, and let it colour also.
Pour in gradually three-quarters of a pint of water and a glassful of sherry ; add a little salt and the bones and trimmings of
the pheasants ; let the sauce boil up, then draw the saucepan
to the side of the fire, and let it keep simmering for an hour
and a half. Strain the gravy and skim carefully, put it back
into the saucepan with the joints of meat, a little saccharin,
and a little lemon-juice. Heat slowly and serve as above.
Time, twenty to thirty minutes to roast the pheasant ; a
quarter of an hour in the first instance, or an hour and a
half in the second, to simmer the sauce.

Pheasant, Roast.

248. Pluck, draw and singe a brace of pheasants. Wipe them with a dry cloth, truss them firmly, and either lard or tie round the breasts a piece of fat bacon. Flour them well, put them before a clear fire, and baste liberally. When they are done enough remove the bacon, serve the birds on a hot dish, and garnish with watercress. Send good brown gravy to table with them. If the fashion is liked, half a dozen of the best of the tail feathers may be stuck into the bird when it is dished. Time, three-quarters of an hour to roast a good-sized pheasant. The drumsticks are generally excellent when devilled.

Peafowl, Larded and Roasted.

249. Choose a young bird, lard it closely over the breast and legs, fill it with a good veal forcemeat—but the forcemeat may be omitted—truss it firmly, and roast before a clear fire for an hour or an hour and a half, according to the size of the bird. When done enough, take off the buttered paper which was round the head, trim the feathers, glaze the larding, and serve the bird on a hot dish with a clear brown gravy under it. Garnish the dish with watercress.

Partridges, Broiled.

250. Prepare the partridges as if for roasting, cut off their heads, split them entirely up the back, and flatten the breast-bones a little. Wipe them thoroughly inside and out with a damp cloth, season with salt and cayenne, and broil over a gentle fire. As soon as they are done enough rub them quickly over with butter, and send them to table on a hot dish, with brown gravy or mushroom sauce in a tureen. Time, fifteen minutes to broil the partridges.

Partridges, Salmi of.

251. Prepare three partridges, lard the breasts well, and roast them, but leave them rather underdone. Leave till

cold ; take off the skin and cut in joints ; put them into a stewpan with over half a pint of good broth, add two or three
shallots and a bit of thin lemon-peel, pepper and salt to taste,
and four teaspoonfuls of Worcester or any other good sauce.
Put it on the fire, and let it stew down to half the quantity.
Strain the sauce through a fine sieve, dish the partridges with
a thin slice of fried bread between the pieces ; pour the sauce
over, and add a squeeze of lemon-juice.

SALMI OF PARTRIDGES, COLD.

252. Prepare as above. When done, strain the sauce and
leave all to become cold. The sauce can have a little Nelson's
gelatine put to it, and be left to set. After it is all nearly cold,
arrange the pieces of partridge in a mould, first putting a little
of the sauce at the bottom, fill up with the sauce, and ice the
whole together. Turn out ; serve with savoury or aspic jelly
round.

PIGEONS, COMPÔTE OF.

253. Truss a dozen plump young pigeons as if for boiling.
Lard them down the breasts, or, if preferred, cover their breasts
with thin slices of fat bacon. Fry them in hot butter till they
are equally and lightly browned all over, then divide them and
put them in side by side in a saucepan large enough to contain
them. Barely cover them with good gravy, and add half a
dozen small onions, a dozen button mushrooms, a glassful of
claret, a little salt and cayenne. Let the birds stew gently for
half an hour, then add a large tablespoonful of tomato sauce,
and stew a few minutes longer. Place the birds on a hot dish,
with the sauce around them.

PIGEONS SERVED WITH WATERCRESS.

254. Roast a couple of young pigeons in the usual way.
Wash and pick two or three bunches of young watercress, and
dry them well ; to do this put them into a dry cloth, take

hold of this by the four corners and shake the leaves until they
are dry ; put them on a dish, sprinkle a little salt over them, lay
the pigeons upon them, and pour brown gravy over. The
cresses are sometimes arranged round the dish instead of being
placed under the birds. Time, about twenty minutes to roast
the pigeons.

MUTTON, NECK OF, BOILED.

255. Shorten the ribs and saw off the chine-bone of a neck
of mutton, or from three or four pounds of the best end ; to
look well it should not exceed five inches in length. Pare off
the fat that is in excess of what may be eaten, and boil slowly
in plenty of water, slightly salted ; skim carefully and remove
the fat from the surface. The meat may be served plainly
with caper or parsley sauce, and a garnish of boiled turnips
and carrots cut into thin strips placed alternately round
the dish. Four middle-sized turnips or three carrots may
be boiled with the mutton. Time, a full quarter of an hour
to the pound.

MUTTON, KEBOBBED.

256. This favourite Oriental dish can be prepared with our
English mutton in a manner far superior to any kebob at
Turkish or Egyptian tables. Take a loin of mutton, joint it
well at every bone, cut off all superfluous fat, particularly of
the kidney, and remove the skin. Prepare a well-proportioned
and large seasoning of the following ingredients : Some
breadcrumbs, sweet herbs, nutmegs, pepper and salt. Brush
the mutton chops over with yolk of egg, and sprinkle the
above mixture thickly over them ; then tie the chops together
in their original order, run a slender spit through them, and
roast before a quick fire, basting them well with butter and
the drippings from the meat, and throwing more of the seasoning on them from time to time. Serve with the gravy from
the meat.

MUTTON, ROEBUCK FASHION.

257. Take a loin of mutton that has been well hung. Remove the fillet, skin, and cut away the fat and bones. Lay the loin in a marinade composed of equal parts of vinegar and water, to a pint of which add a glass of port or claret, a couple of carrots, and two large onions cut into quarters with a clove in each, a dozen peppercorns, two blades of mace, a bunch of herbs and parsley, some bay-leaves, and two tea-spoonfuls of salt. When the mutton has lain in the marinade twenty-four hours turn it, and let it lie until next day; then drain, and put it into a braising-pan with a little of the pickle, the pan being well lined with bacon. Stew it three hours. Glaze the meat and serve with gravy, adding walnut ketchup and a glass of claret.

INDIAN FAGADU.

258. Pick the meat from a lobster and a pint of shrimps, cut it into small bits, and season it with an onion and a clove of garlic shred finely, and some cayenne and salt. Prepare some spinach as for boiling, put it into a stewpan in the usual way without water, add the lobster, and stew gently with an onion or two sliced and previously fried in butter, keeping the lid closed for some time. When nearly done, stir the contents over the fire to absorb the moisture, and when quite dry queeze in some lemon-juice.

CURRIED SWEETBREADS.

259. Wash and soak one or two sweetbreads, put them into some boiling water with a little salt, an onion, and a bit of parsley; let them simmer for ten minutes (any pale broth is better than water); take them out, leave them to drain and get quite cold, then cut them into slices, about half an inch thick, and fry lightly. Have some good curry, put the slices into this, and stew gently for twenty minutes; cocoanut-milk must be added with the sweetbread, **or** about ten minutes before serving.

LEVERET, ROASTED.

260. Leverets may be used when hares are out of season. They should be trussed in the same way, and may be stuffed or not (with hare stuffing), according to preference. A leveret is best when larded, but if this cannot be done, cover it either with thin slices of fat bacon or with a thickly buttered piece of white paper. Roast it before a brisk fire, and baste it constantly, and a few minutes before it is taken down remove the bacon or paper. Serve it very hot, and send red-currant jelly to table with it as well as the following gravy, a little of which may be put in the dish and the rest in a tureen : Thicken half a pint of stock with a small piece of butter rolled in flour, let it boil for ten minutes, then stir a wineglassful of port into it, boil up once more, and serve. Time, an hour to roast the leveret.

MEAT PIE À LA DON PEDRO.

261. This is a kind of ragoût put into a tin made expressly for the dish. Take some mutton chops, either from the loin or neck, trim them neatly, and toss them with some chopped parsley, butter, pepper, and salt, etc., in a stewpan over a slow fire. Place the chops with some good brown stock into the tin baking-dish, and add slices of lean ham, and cover with the lid.

OMELETTE WITH GRAVY.

262. Whisk half a dozen fresh eggs thoroughly, and mix with them a small pinch of salt, two pinches of pepper, a tablespoonful of finely minced parsley, half a teaspoonful of chopped onions, and two tablespoonfuls of nicely seasoned gravy. Dissolve two ounces of fresh butter in a hot frying-pan, over a gentle fire, and fry the omelette in the usual way. Serve it on a hot dish with half a pint of good gravy poured round it. Time to fry, six or seven minutes.

Onions with Beefsteak, etc.

263. Take two large Spanish onions, remove a thin piece off each end, peel off the outer skins, and cut them into slices a quarter of an inch thick. Place an ounce of butter or good dripping in a saucepan, let it melt, then put with it a pound of steak, dividing it into pieces a little thinner than for boiling. Brown these in the butter, add a little pepper and salt, the sliced onions, three ounces more of butter, but no liquid; cover the saucepan closely, and simmer as gently as possible till done. Arrange the steak neatly in the centre of a hot dish, boil up the onion gravy sauce with a tablespoonful of walnut ketchup, pour it over the meat, and serve immediately. Chickens or rabbits are sometimes cooked in the same way. Time, about an hour and a half.

Filets de Bœuf aux Truffes.

264. Cut out the inside of a sirloin of beef, beat it well to make it tender, cut it in slices, trimming them neatly; lay them in oil, and let them soak for ten minutes, then fry in butter. Chop up some parsley, lemon thyme, half a shallot, and slice some truffles (that have been previously cleaned and brushed, boiled for twenty minutes in some good stock, quarter of a pint, and half a pint of white wine, pepper and salt), add fifteen drops of vinegar. Lay the herbs in the middle of the dish and the fillets round, and the truffles round the fillets.

Boiled Rabbits.

265. Select very young ones for boiling; wash and clean well. Fasten the head to the side. Have water boiling and skimmed ready, put in the rabbits, and simmer gently for half to three-quarters of an hour.

Roast Rabbit.

266. Rabbits can be cooked much the same as hares, but they will take hardly an hour to roast. If liked, the backbone can be taken out; it must be carefully done, so as not to break the skin. A rabbit done this way will require much more stuffing put into it; a little thin bacon or slices of ham may be put in before the stuffing. Baste well. Red-currant jelly is also sent in with roast rabbit. (See GRAVY.)

RABBIT À LA TARTARE.

267. Take a rabbit and bone it. Then cut it into pieces, which marinade some hours in parsley, mushrooms, chives and a clove of garlic, all chopped fine, together with pepper, salt and oil; dip each piece of rabbit in breadcrumbs, broil, sprinkling the pieces with the marinade. Serve in a sauce à la Tartare. (See TARTAR SAUCE.)

RABBIT WITH TOMATO SAUCE.

268. Cut a rabbit in pieces, fry a light brown slowly (it is best done in a stewpan) in a little butter or lard, add pepper, salt and a small sliced onion. Pour on to it some tomato sauce, a few spoonfuls of gravy, and stew for half to three-quarters of an hour. Rabbits that have been roasted and are rather underdone can be cooked this way after.

STEWED RABBIT.

269. Cut up into neat joints. Take a large flat stewpan, make it hot, put into it a lump of lard about the size of a large walnut, put in the joints of rabbit so that each piece touches the bottom of the pan, add a little sliced onion, and sprinkle over them a little flour; brown both sides. Then pour over it a pint of boiling water or weak stock, season with salt, pepper, and two or three cloves, a bit of allspice, a bunch of parsley, and two bay-leaves. Simmer slowly for one hour and a half, then add a tumbler of claret, and simmer again for one hour. It must not be allowed to do fast, or it will dry up too much.

RABBIT, FRICASSEED.

270. Cut a young rabbit into neat joints, lay it in a stew-pan, and cover with good stock. Let the liquid boil, then put

with it three onions, three carrots, three turnips, and three
sticks of celery, all sliced; add a bunch of parsley, a sprig of
thyme, a blade of mace, a saltspoonful of grated nutmeg, and
a small quantity of saccharin, and stew all gently together
until the vegetables are quite soft. Lift the vegetables out,
rub them through a sieve with the back of a wooden spoon.
Stir the purée over the fire with a tablespoonful of the gravy for
two or three minutes to make it quite hot. Put the pieces of
rabbit on a dish, cover with the purée, and pour the sauce
over all. Serve very hot. Time, one hour.

RABBIT, RAGOÛT OF.

271. Skin, empty and wash a plump young rabbit, cut it
up into ten or twelve pieces, and lay it in a saucepan with a
dozen button mushrooms, half a dozen small onions, a bunch
of parsley, a sprig of thyme, and a bay-leaf. Pour over these
ingredients as much boiling stock or water as will cover them,
and let them simmer very gently until the rabbit is tender.
Lift out the rabbit, skim and strain the sauce, and thicken
with a tablespoonful of brown thickening. Season with salt,
pepper and grated nutmeg, and let it boil till smooth. Add a
wine-glassful of chablis, if liked. Put in the pieces of meat.
Let them get thoroughly hot without allowing the gravy
to boil, arrange them neatly in a dish, pour the gravy over
them, and serve very hot. Time to simmer the rabbit, from
an hour and a half to two hours.

CHICKEN À LA MARENGO.

272. Cut a fine chicken into neat joints, season it with salt
and cayenne, and fry it till done in about half a tumblerful of
oil or clarified butter. When half cooked, add a clove of
garlic, two shallots, and a faggot of sweet herbs. Drain the
meat from the fat, and mix with the latter a teaspoonful of
flour, and very gradually sufficient good stock to make the
sauce of the consistence of thick cream. Stir it till it is
thick and smooth. Put the chicken on a hot dish, strain the

sauce over it, and serve. If liked, mushrooms or fried eggs may be taken to garnish the dish. Time, about twenty-five minutes to fry the chicken.

Chicken, Devilled.

273. The best parts of chicken for a devil are the wings and legs. Remove the skin, score the flesh deeply in several places, and rub in a mixture made of salt, pepper, cayenne, mustard, anchovy and butter. Broil over a clear fire, and serve the fowl hot on a napkin.

Chicken, Fried.

274. Take the remains of a cold chicken, cut it into neat pieces, brush a little oil over each piece, and strew over it rather thickly salt and curry powder. Melt a little butter in a frying-pan, and fry some onions, cut into thin strips about half an inch long and the eighth of an inch wide ; fry them slowly, and keep them in the pan until they are a dark brown colour and quite dry. They will require a little care, as they must on no account be burnt. Fry the chicken, strew the onions over it, and serve with slices of lemon. Time to fry the chicken, ten minutes.

Calf's Liver à la Mode.

275. After well washing the calf's liver, soak it for a short time in cold water, then wipe it dry, and insert lardoons of bacon at equal distances in the interior part of the liver ; put it into a stewpan with about two ounces and a half of butter, a small bunch of sweet herbs tied together, half a blade of mace, and a small onion stuck with six cloves, and fry it a nice brown ; then add three carrots, two turnips, an onion cut into wheels, and a wineglassful of brandy with sufficient water to just cover the whole. Baste it frequently with its own gravy, and let it simmer slowly for two hours. When done, take out the liver and put it on a dish garnished with the cut vegetables ; strain and skim the gravy ; add one tablespoonful of Harvey-

sauce and a glass of wine; boil it to the quantity required. Pour it over the liver and serve it up hot.

BRAISED LOIN OF LAMB.

276. Bone a loin of lamb and line the bottom of a stewpan, just capable of holding it, with a few thin slices of fat bacon; add to it one bunch of green onions, five or six young carrots, a bunch of savoury herbs, two blades of pounded mace; cover the meat with a few more slices of bacon, pour in a pint of stock, and simmer very gently for two hours. Take it up, dry it, strain, and reduce the gravy to a glaze, with which glaze the meat, and serve it either on stewed spinach or stewed cucumbers.

CURRIED BEEF.

277. Cut up some beef into pieces about one inch square, put a little butter into a stewpan with a little onion sliced, and fry them of a light brown colour; add one dessertspoonful of curry-powder, quarter of a pint of stock or gravy, and stir gently over a brisk fire for about ten minutes. Should this be thought too dry, a spoonful or two more of gravy may be added; but a good curry should not be very thin. Serve with sippets of well-toasted bread. A nice way of doing up cold beef.

TO DRESS BEEF KIDNEYS.

278. Cut a beef-kidney into neat slices, put them into warm water to soak for two hours, and change the water two or three times; then put them on a clean cloth to dry the water from them, and lay them in a frying-pan with some clarified butter, and fry them of a nice brown; season each side with pepper and salt, put them into a stew-pan, and then gently stew for an hour. Put them round the dish, and the gravy in the middle. Before pouring the gravy in the dish add one tablespoonful of lemon-juice and a very small quantity of saccharin.

Minced Beef.

279. Put into a stew-pan a little butter with an onion chopped fine; add a little gravy and one tablespoonful of strong ale; season with pepper and salt, and stir these ingredients over the fire until the onion is a rich brown. Cut, but do not chop, some cold beef *very fine*, add it to the gravy, stir till quite hot and serve; garnish with sippets of well-toasted bread. Be careful in not allowing the gravy to boil after the meat is added, as it would render it hard and tough.

To Roast Venison.

280. Take great care the meat is well-hung. Take the skin from the top part, put butter and salt over the fat, then make a paste of flour and water, lay it on, and fasten it by four or five sheets of paper, sewn together and skewered on firmly. Roast it gently and baste it. About half an hour before it is done take off the paper and paste, and let it colour gradually to a pale brown, but be most careful it does not burn. Serve with a good gravy round, and with warmed red-currant jelly. A little wine can be added to the gravy.

Hashed Venison.

281. Take a pint of very good brown gravy, mix with it some of the sauce Tartare. Cut the venison in small, thin slices of equal size, put them into a saucepan, cover with the sauce, and let it stand from fifteen to twenty minutes; put the pan on the fire, let it get gradually warm till it gets hot quite through. It must on no account boil, but it must be very hot when sent to table.

Norman Hash.

282. Take two ounces of butter, warm it, add half an onion, a sprig of herbs, and a good pint of stock ; boil well and strain; slice mutton very thin and free from skin ; put the meat into the sauce, stew very gently for three-quarters of an hour, take out the herbs. Add half a glass of sherry before serving.

Beef can also be used for this hash, but will require rather longer to stew. Cooked meat is best for this.

To Stew Mutton Cutlets (Plain).

283. These can be taken from either a loin or neck of mutton ; free them from skin and fat. Fry slightly, either plain or cover with egg and breadcrumbs. Have a good cold gravy ready ; put in your cutlets, cover your pan, and let them stew gently for an hour. Add a few button mushrooms to the gravy before serving.

Calf's Head Stewed with Oysters.

284. Soak half a small calf's head (without the skin) for one hour in cold water with a teacupful of vinegar in it. Well wash it in two or three waters, put it into a stewpan with two onions, a bay-leaf, a laurel-leaf, a sprig of thyme, a sprig of marjoram, two sage leaves, four sprigs of parsley, two cloves, four allspice, six black peppercorns, half a carrot, and a pint and a half of cold water. Boil up quickly, skim, then simmer gently for an hour and a half, skimming constantly. Take out the head, strain the liquor, add to it one tablespoonful of baked flour and the strained liquor of three dozen oysters ; boil up, put the head in again and continue to simmer for three quarters of an hour longer ; add three dozen oysters ; let it just simmer again and serve. It must not boil after the oysters are added.

Calf's Head Ragoût.

285. Wash half a calf's head thoroughly, and boil it for about three hours ; take it up, drain it, and score the outside skin in diamonds. Brush it over with well-beaten egg, and strew over that a few finely-grated breadcrumbs, a tablespoonful of chopped parsley, a teaspoonful of chopped thyme, a teaspoonful of salt, and half a saltspoonful of cayenne. Put it in a hot oven, or place it before the fire to brown, and before sending

it to table squeeze over it the juice of a large lemon, and a little oiled butter poured over it.

LAMB CHOPS WITH CUCUMBER SAUCE.

286. Dip the chops in beaten egg and brown breadcrumbs, and fry them. When nicely browned, arrange them in a circle on a hot dish and put in the centre a sauce prepared as follows: Peel a young fresh cucumber, and cut it into dice; strew a little pepper and salt over these. Melt three or four ounces of butter in a saucepan, put in the cucumber, cover it closely and place it on a moderate fire, shaking the pan frequently to prevent sticking. When it is steamed until the pieces of cucumber are quite tender but unbroken, serve them in the centre of the dish. It will take about ten minutes to fry the chops, and about twenty minutes to stew the cucumber.

VEGETABLES.

The following vegetables are suitable for corpulent people during the respective months given.

JANUARY.

Asparagus (forced), Jerusalem artichokes, broccoli, Brussels sprouts, cabbages, cardoons, celery, chervil, cresses, endive, lettuces, savoys, Scotch kale, spinach, turnips, herbs.

FEBRUARY.

Jerusalem artichokes, asparagus (forced), broccoli, brussels sprouts, beans (French or kidney), cabbages, celery, cardoons, chervil, cresses, cucumbers (forced), endive, lettuces, savoys, spinach, seakale, turnips, various herbs.

MARCH.

Artichokes (French), asparagus (forced), broccoli, Brussels sprouts, beans (forced), cabbages, celery, chervil, cresses, cucumber (forced), endive, kidney beans, lettuces, radishes (early), savoys, seakale, spinach, turnips, turnip-tops, various herbs.

APRIL.

Artichokes (French), asparagus, beetroot, beans (French and kidney, forced), broccoli, celery, chervil, cucumbers (forced), lettuce, cabbages, radishes, young onions, small salad, seakale, spinach, sprouts, turnip-tops, various herbs.

MAY.

Artichokes (French), asparagus, beans, cabbages, chervil, cucumbers, cauliflower, cresses, lettuces, peas, radishes, salad, seakale, spinach, turnip-tops, and various herbs.

JUNE.

Artichokes, asparagus, beans, cabbages, chervil, cucumbers, cauliflower, endive, lettuces, onions, peas, radishes, small salad, seakale, sorrel, spinach, turnips, various herbs.

JULY.

Artichokes, beans, cabbages, cauliflowers, cucumbers, cresses, endive, lettuces, mushrooms, onions, peas, radishes, red cabbages, small salads, salsify, seakale, sorrel, spinach, sprouts, turnips, vegetable marrows, various herbs.

AUGUST.

Artichokes, beans, cabbages, cauliflowers, cucumbers, cresses, shallots, endive, lettuces, mushrooms, onions, peas, radishes, red cabbages, seakale, small salads, salsify, sprouts, turnips, vegetable marrows, and various herbs.

SEPTEMBER.

Artichokes, Jerusalem artichokes, beans, cabbage, sprouts, cauliflower, celery, carrots, endive, shallots, lettuces, leeks, mushrooms, onions, peas, salads, seakale, sprouts, tomatoes, turnips, vegetable marrows, various herbs.

OCTOBER.

Artichokes, Jerusalem artichokes, beets, broccoli, cabbages, cauliflower, celery, cucumbers, endive, shallots, lettuces, leeks, mushrooms, onions, sprouts, tomatoes, turnips, vegetable marrows, various herbs.

NOVEMBER.

Jerusalem artichokes, broccoli, cabbages, carrots, celery, cardoons, endive, leeks, onions, salad, spinach, sprouts, Scotch kale, various herbs.

DECEMBER.

Jerusalem artichokes, beetroot, broccoli, cabbages, carrots, cardoons, celery, leeks, onions, spinach (winter), Scotch kale, turnips.

CABBAGE, BOILED.

287. Cut off the stalk, remove the failed and outer leaves, and halve, or, if large, quarter the cabbages. Wash them thoroughly and lay them for a few minutes in water, to which a tablespoonful of vinegar has been added, to draw out any insects that may be lodging under the leaves. Drain them in a colander. Have ready a large pan of boiling water, with a tablespoonful of salt and a small piece of soda in it, and let the cabbages boil quickly till tender, leaving the saucepan uncovered. Take them up as soon as they are done, drain them thoroughly and serve. Time to boil young summer cabbages, from ten to fifteen minutes; large cabbages, half an hour or more.

SAVOY CABBAGE.

288. The savoy is a large, close-hearted cabbage, seasonable in winter. It may be dressed according to the instructions already given for boiling cabbages. A savoy cabbage will need to boil thirty minutes or more, according to size.

BROCCOLI, BOILED.

289. Trim off all leaves that are not required or liked, and place the broccoli in a pan of salted water to kill any insects, etc., that may have taken shelter under the stalks. Wash them well and put them into an uncovered saucepan of boiling water, with a large tablespoonful of salt to every half gallon of water. Keep them boiling till done, which will be in about ten or fifteen minutes, according to size. Drain them directly they are done, or they will lose colour and become sodden.

SCOTCH KALE.

290. Like all other greens, Scotch kale should be procured as fresh as possible. Cut away the outer and decayed leaves and the stalks, wash the kale with scrupulous care, and drain it. Put it into boiling water slightly salted, and let it boil quickly until done enough. Take it up, drain it thoroughly, and serve very hot. Whilst the kale is boiling, the saucepan should be left uncovered. Time to boil, twenty minutes.

SEAKALE, BOILED.

291. When fresh and delicately cooked, seakale resembles, and will serve as a substitute for, asparagus. Carefully wash and brush the seakale to remove the sand and grit, cut out the black part of the roots and tie the shoots up in small bundles, and put it into a stewpan of boiling water with a teaspoonful of salt; let it boil for about twenty minutes, or until tender.

SPINACH, BOILED.

292. Take two pailfuls of spinach, young and freshly-gathered, pick away the stalks, wash the leaves in several waters, lift them out with the hands that the sand or grit may remain at the bottom, and drain them on a sieve. Put them into a saucepan with a good sprinkling of salt and the water

which clings to the leaves, and let them boil until tender. Take the spinach up, drain it, and press it well ; chop it small, and put it into a clean saucepan with a little pepper and salt and a slice of fresh butter; stir it well for five minutes. Serve on a hot dish, and garnish with fried sippets. Time to boil the spinach, ten to fifteen minutes.

LETTUCE, STEWED.

293. Take four good-sized lettuces, trim away the outer leaves and the bitter stalks, wash the lettuces carefully, and boil them in plenty of salted water until they are tender. Lift them into a colander, and squeeze the water from them ; chop them slightly, and put them into a clean saucepan with a little pepper and salt and a small piece of butter. Dredge a little flour on them, pour over them three tablespoonfuls of good gravy, and simmer gently for three quarters of an hour, stirring all the time. Squeeze a dessertspoonful of vinegar or lemon-juice upon them, and serve as hot as possible.

MUSHROOMS, GRILLED.

294. Cut the stalks, peel, and score lightly the underside of large mushroom flaps, which should be firm and fresh-gathered ; season them with pepper and salt, and steep them in a marinade of oil or melted butter. If quite sound they may be laid on a gridiron over a slow even fire, and grilled on both sides, but they are best done in the oven if at all bruised. Serve on a hot dish, with a piece of butter on each mushroom and a squeeze of lemon-juice. Time, about twelve minutes to grill ; five minutes to steep in marinade.

CELERY, STEWED.

295. Wash four heads of celery very clean, trim them neatly, cutting off the leaves and tops ; cut them into three-inch lengths and tie them in small bundles, and parboil them in sufficient salt and water to cover them. Drain, and stew them until tender in some stock. Brown two ounces of butter

with a teaspoonful of flour in a saucepan, dilute it with the stock in which the celery was boiled, lay the celery in it, let it boil for ten minutes more, and serve as hot as possible. Time, three-quarters of an hour.

CARDOONS, BOILED.

296. Choose a few heads of sound white cardoons. Cut them into pieces about six inches long, remove the prickles, and blanch them in boiling water for a quarter of an hour. Scrape off the skin, and tie them in bundles. Cover them with nicely-flavoured stock, and boil till tender.

TURNIPS, BOILED.

297. Turnips should only be served whole when they are very young. When they have reached any size they should be mashed. Pare the turnips and wash them; if very young a little of the green top may be left on; if very large they should be divided into halves, or even quarters. Throw them into slightly salted water, and let them boil gently till tender. Drain and serve them. Time to boil old turnips, three-quarters of an hour to an hour and a half; young turnips, fifteen to twenty-five minutes.

ARTICHOKES, BOILED.

298. Gather the artichokes two or three days before they are required for use. Cut off the stems, pull out the strings, and wash them in two or three waters, that no insects may be in them. Have a large saucepan of boiling water, with two tablespoonfuls of salt and a piece of soda the size of a sixpence to every gallon of water, put the artichokes in with the tops downwards, and let them boil quickly until tender. About half an hour or three-quarters will boil them, but that can be ascertained by pulling out one of the leaves (if it comes out easily they are done), or by trying them with a fork. Take them out and lay them upside down to drain. Serve them on a napkin.

Asparagus, Boiled.

299. Scrape very clean all the white part of the stalks from the asparagus, and throw them into cold spring water, tie them up in bundles, cut the root ends even, and put them in a piece of muslin to preserve the tops. Have a wide stewpan of spring water, with one tablespoonful of salt to half a gallon of water, and when it boils lay in the asparagus and boil it quickly for fifteen minutes, or until it is tender. Lay them in the dish with the white ends outwards and the points meeting in the centre.

To Boil French Beans.

300. Take as many French beans as you may require, cut off the tops and bottoms, and remove the strings from each side; then divide each bean into three or four pieces, cutting them lengthways, and as they are cut put them into cold water with a little salt. Have ready a saucepan of boiling water, drain the beans from the cold water, and put them in. Boil them quickly with the saucepan uncovered, and as soon as they are done drain them in a colander. Dish, and serve them with a small piece of butter stirred into them.

Beans, Stewed.

301. Have ready a good rich brown gravy. Cut up some small onions, chives, and parsley; throw them into the gravy, and simmer for ten minutes before the beans are put in. Sprinkle a quart of beans with two teaspoonfuls of salt, one of pepper, and a small quantity of saccharin; mix together, and put them into the gravy. Stir the beans gently over a slow fire till the gravy is absorbed by them. In ten minutes serve them up.

Cauliflowers, Boiled.

302. Make choice of some cauliflowers that are close and white, pick off all the decayed leaves, and cut the stalk off

flat at the bottom ; then put them with the heads downwards
in strong salt and water for an hour, to draw out all the
insects. Drain them in a colander, and put them into a saucepan with plenty of fast boiling water; keep the pan uncovered, and boil them quickly until tender, which will be
from twelve to fifteen minutes, or longer if they are very
large. Skim the water clean, and when done take them up
with a slice and serve.

SQUASH, OR VEGETABLE MARROW, BOILED.

303. Peel the marrows, and put them into a saucepan of
boiling water and salt (one tablespoonful of salt to half a
gallon of water). When tender, take them out; cut them
into quarters if large ; if not, halve them. Serve them in a
vegetable dish.

SQUASH, OR VEGETABLE MARROW, IN GRAVY.

304. Boil a large marrow in the usual way. When three-
parts cooked, take it up, cut it into squares, place these in a
saucepan, and pour over them as much thick brown gravy as
will cover them. Let them heat gently. Serve in a vegetable
dish with the gravy poured round them.

GREEN PEAS, BOILED.

305. Shell half a peck of green peas, and put them into a
saucepan of boiling water with a teaspoonful of salt and a
sprig or two of mint ; let them boil about half an hour with
the pan closely covered. When tender, drain them through a
colander, and put them in a dish with a bit of butter stirred
into them. Serve them up very hot.

LEEKS, BOILED.

306. Leeks are generally used in soups, etc. If served
alone, take them when very young, trim off the root, the
outer leaves and the green ends, and cut the stalks into six-
inch lengths. Tie them in bundles, put them into boiling

water, with a dessertspoonful of salt and a tablespoonful of
vinegar, and let them boil until quite tender. Drain them
and serve.

BRUSSELS SPROUTS.

307. Pick, trim, and wash a number of sprouts; put them
into plenty of fast boiling water. The sudden immersion of
the vegetables will check the boiling for some little time, but
they must be brought to a boil as quickly as possible, that
they may not lose their green colour; add a tablespoonful of
salt, keep the saucepan uncovered, and boil very fast for fifteen
minutes. Lose no time in draining them when sufficiently
done.

TOMATOES.

308. Cut in slices, fry in butter just brown; add one
tablespoonful of white vinegar, chili, a few drops of tarragon,
one saltspoon of salt, and a little saccharin. Simmer twenty
minutes.

BROILED MUSHROOMS.

309. Skin the mushrooms and cut off the stalks; put them
in a Dutch oven in front of the fire, with a little butter,
pepper, and salt. Serve on toast thinly buttered.

MASHED TURNIPS.

310. Take six moderate-sized turnips, pare them neatly, and
put them into cold water to blanch for half an hour; then
put them into boiling water, and boil about half an hour;
drain and press out all the water, and rub the turnips through
a wire sieve; put them into a stewpan with half a gill of
thick cream and a saltspoonful of salt; stir till boiling hot,
then serve.

BOILED ONIONS.

311. Peel the onions, and boil them in salt and water for
ten minutes; throw them into cold water for half an hour,

then put them into a saucepan, and well cover them with cold water, and let them boil gently for an hour. Drain, and serve with or without dissolved butter over them.

PORTUGAL ONIONS, STEWED.

312. Peel the onions, and place them in a stewpan; for each onion knead together half an ounce of butter and a little saccharin; put it on the onions, and let them slowly become slightly browned. Then pour over each a teaspoonful of tomato sauce and a tablespoonful of gravy or stock: simmer gently for three hours, basting the onions frequently with the gravy. Serve very hot.

PORTUGAL ONIONS, ROASTED.

313. Peel the onions and place them in a Dutch oven before a good fire, baste them frequently with dissolved butter (an ounce for each), and roast for an hour and a half. Serve with or without their own sauce.

PORTUGAL ONIONS, CURRIED.

314. Peel and wash the onions, put them into a saucepan with plenty of water and a little salt. Boil (uncovered) till tender; then press out the water and chop; put them into an enamelled saucepan with a little butter and a little curry powder. Simmer for five minutes and serve.

RAGOÛT OF CELERY.

315. Wash well, and boil half an hour. Take out the celery, put it into cold water for a quarter of an hour, then strain well. Stew in good gravy with a little mushroom ketchup, salt, and pepper. Serve hot.

A MIXED VEGETABLE.

316. Peel an onion, slice and fry in butter, then stew in a little broth till tender; have some turnips, carrots, and celery boiled, and cut in neat pieces. Put them to the onion, with salt, pepper, and a little mustard (half a teaspoonful of

French mustard is the best for this. Simmer till hot, and serve.

FRIED CABBAGE.

317. Boil the cabbage in two waters and skim well, drain very well, chop or slice it, fry in butter or good dripping, with pepper and salt. Serve on a hot dish with a few slices of fried beef or ham over it.

SPINACH WITH CREAM.

318. Boil and drain two pounds of spinach in the usual way. Press it between two plates to free it thoroughly from moisture, and heat it in a clean saucepan with a little pepper and salt and a small lump of butter. When it is dry, add very gradually two tablespoonfuls of boiling cream, and simmer it gently for five minutes. Serve very hot.

SPINACH WITH GRAVY.

319. Prepare the spinach in the usual way as in the foregoing recipe. Dissolve two ounces of fresh butter in a saucepan, put in the spinach and stir it till the butter has dried away. Add a teaspoonful of salt, a very little saccharin and as much grated nutmeg as will cover a sixpence. Stir it again, and moisten with two tablespoonfuls of highly-seasoned veal broth and a teaspoonful of chilli vinegar. Stir it over the fire till the liquid is absorbed, and serve very hot.

CABBAGE, CREAMED.

320. Thoroughly cleanse two young cabbages and boil them until quite soft. Take them out, drain and press them between two hot plates until they are dry, when they may be slightly chopped. Melt a piece of butter the size of an egg in a stewpan, add pepper and salt, then put in the cabbage and turn it about for two or three minutes. When it is thoroughly heated, mix with it very gradually a cupful of cream.

ARTICHOKES, FRIED.

321. Pare some artichokes, and boil them in salt and water for about a quarter of an hour. Drain and cut them into slices about a quarter of an inch in thickness, dip them into the white of an egg well beaten, and afterwards strew finely-grated bread on them. Fry in boiling oil or lard till they are nicely browned, and serve piled high on a dish.

ARTICHOKES STEWED IN GRAVY.

322. Strip off the leaves from the artichokes, remove the chokes, and soak them in lukewarm water for three hours, changing the water three or four times. Place them in a saucepan with enough gravy to cover them, a tablespoonful of mushroom ketchup, the juice of a lemon, and a piece of butter the size of a walnut. Let them stew gently until tender, then serve with the sauce poured over them, and as hot as possible.

ARTICHOKES, MASHED.

323. Wash and pare some artichokes ; boil them in salt and water until quite tender, then drain and press the water thoroughly from them. Put them into a saucepan, and beat to a pulp, adding salt, pepper, and a little cream. Serve very hot.

STEWED RED CABBAGE.

324. Cut a cabbage in shreds, wash it well in salt and water, put it into a stewpan without draining it much, add pepper and a little broth, a tablespoonful of vinegar, and a small lump of butter. Stew till tender.

STEWED CUCUMBERS.

325. Peel and core the cucumbers, cut them into neat pieces, fry in a little butter, put the fried pieces into a stewpan with a little good gravy, pepper, salt, and a teaspoonful of vinegar. Stew gently till tender.

HARICOTS VERTS.

326. Cut French beans very thin (they should be young and
tender), boil and drain well; add salt, pepper, a little good
gravy, a bit of butter; shake well over the fire. Take them
off the fire, and stir-in the beaten yolk of an egg and a little
lemon-juice. Serve very hot.

CURRIED TOMATOES.

327. Cut the tomatoes in slices; either bake or fry them
lightly. Grate an apple, chop a bit of onion small, fry in hot
butter till quite tender, add two large teaspoonfuls of curry
powder, put in a few spoonfuls of good thick gravy, simmer a
few minutes, add the tomatoes with a very little lemon-juice;
let it be rather thick. Serve hot.

SEAKALE, STEWED.

328. Wash the seakale, and tie it in bundles. Boil it in
salted water for a quarter of an hour, then drain it, and put it
into a saucepan with as much brown gravy as will cover it;
stew gently till tender. Lay it in a hot dish, stir a little
lemon-juice into the sauce and pour it over.

CAULIFLOWER WITH SAUCE.

329. Boil two large white cauliflowers in a little salt and
water until tender, then cut off the stalks and press them head
downwards into a hot basin. Turn them into a tureen, and
pour round them a little tomato or piquante sauce.

SQUASHES, OR VEGETABLE MARROWS, FRIED.

330. Boil the marrows in the ordinary way till they are
tender but quite firm. Let them get cold, and cut them into
slices. Brush them over with egg, dip in finely-grated breadcrumbs, and fry till they are lightly browned.

Squashes, or Vegetable Marrows, Mashed.

331. Boil two good-sized vegetable marrows in a little salt and water till tender. Take them up, drain them, turn them into a bowl and mash them with a wooden spoon. Heat them in a saucepan with a piece of melted butter the size of a walnut, and a little pepper and salt. Marrows dressed thus are excellent, served piled high in the centre of a dish of cutlets.

Tomatoes, Baked.

332. Slice six or eight ripe tomatoes, season with pepper and salt, and sprinkle breadcrumbs lightly over them; divide about two ounces of fresh butter into little pieces, and place these here and there upon them. Bake in a moderate oven. Serve on a hot dish as an accompaniment to roast meat of all kinds.

Stewed Mushrooms.

333. Take off the skin and stems, wash the mushrooms quickly, place them in a stewpan (an earthen one is best) with two ounces of butter, a tablespoonful of water, a teaspoonful of vinegar, a saltspoonful of pepper, a teaspoonful of salt. Simmer for twenty minutes, throw in half a gill of cream, and serve very hot.

Brussels Sprouts, Sauté.

334. Wash and drain one pound of sprouts; put them into boiling salt and water for a quarter of an hour. When done, dry them on a clean cloth; dissolve half an ounce of butter in a pan, and shake the sprouts in it over the fire for a minute or two; season them with pepper, salt, and a little nutmeg, and serve very hot. Sprouts about the size of a walnut have the most delicate flavour.

SACCHARIN.

The invention of the above substitute for sugar is a boon to those who suffer from corpulency or skin disease, as the use of sugar in both cases is injurious, and this article, for which we are indebted to Dr. Fahlberg, is a perfect substitute. Experiments have been made of the most exhaustive nature, that prove it to be perfectly harmless, and Dr. Pavy and others give the following as the result of their investigation :

(1) That saccharin is quite innocuous when taken in ordinary dietary.

(2) Saccharin does not interfere with or impede the digestive processes when taken in any ordinary quantity. The *Lancet* says its continued use is quite harmless.

This being so, sugar—one of the great dietetic articles so constantly used in food, and the greatest of all fattening substances—may be completely cast aside, as saccharin is a perfect substitute, and quite as convenient to use. It is sold in a soluble powder and in the form of a tabloid, containing half a grain, this being equal to about half an ounce of sugar.

For stewed fruits, jellies, and all culinary purposes it will be found that two tabloids are equal to quite one ounce of sugar. They are soluble in hot or cold water. One tabloid, or half a grain of the powder, is sufficient for a large cup of tea or coffee. The tabloids are sold at about 25 cents a hundred. The best-known English makers are Burroughs, Wellcome and Co., Snow Hill Buildings, London, and they may be had of all leading druggists, grocers, and chemists in the larger American cities.

The recipes for jellies, beverages, etc., given in this book have been tried by the author, and he can speak of them as in every way equal to those containing sugar. If it is desirable to make a *firm* jelly, a little more isinglass or gelatine should be used than is given in the recipe. Sugar, as is well known, is the most fattening article in the daily dietary, and with some people the most bilious, so that it is equally a boon to those who are troubled with an inactive liver. This article has a great future before it, and should entirely take the place of sugar in the dietary of those disposed to corpulency. By its aid stewed fruits, tea, coffee, and other daily requisites are made palatable.

FRUITS, JELLIES, AND CREAMS, SWEETENED WITH SACCHARIN.

LEMON CREAM.

335. Pare into a pint of water the peel of three large lemons ; let it stand four or five hours ; then take them out, and put to the water the juice of four lemons and four grains of saccharin, or eight tabloids dissolved in a little boiling water. Beat the whites of six eggs, and mix it altogether ; strain it through a lawn sieve, set it over a slow fire, stir it one way until as thick as good cream, then take it off the fire, and stir it until cold, and put it into a glass dish. Orange cream may be made in the same way, adding the yolks of three eggs.

LEMON CREAM WITHOUT CREAM.

336. Put a quart of new milk into a stewpan with the peel of three small lemons cut thin, four grains of saccharin or eight tabloids, three-quarters of an ounce of bitter almonds, blanched and pounded to a paste, and about two ounces of gelatine or isinglass. Boil the whole over a moderate fire for eight or nine minutes, until the gelatine or isinglass is thoroughly dissolved ; then strain it through a fine sieve into a jug with a lip to it, stir in the yolks of seven well-beaten eggs, and pour the mixture from one jug to another until barely cold ; then add the strained juice of three small lemons, stir it quickly together, and pour it into an oiled mould.

Raspberry Cream without Cream.

337. Mix with a quarter of a pound of raspberries three grains of saccharin or six tabloids, and the whites of four eggs. All to be beaten together for one hour, and then put in lumps in a glass dish.

Calf's-foot Jelly.

338. Cut two calves' feet in small pieces after they have been well cleaned and the hair taken off. Stew them very gently in two quarts of water till it is reduced to one quart. When cold, take off the fat and remove the jelly from the sediment. Put it into a saucepan with six grains of saccharin or twelve tabloids, a pint of white wine, a wineglass of brandy in it, the peel of four lemons finely chopped, the whites of four eggs well-beaten, and their shells broken. Put the saucepan on the fire, but do not stir the jelly after it begins to warm. Let it boil a quarter of an hour after it rises to a head ; then cover it close, and let it stand about half an hour; after which, pour it through a jelly-bag, first dipping the bag in hot water to prevent waste, and squeezing it quite dry. Pour the jelly through and through until clear, then put it into a mould.

Gooseberry Fool.

339. Put two quarts of gooseberries into a stewpan with a quart of water; when they begin to turn yellow and swell, drain the water from them and press them with the back of a spoon through a colander; sweeten them with saccharin to your taste, and set them to cool. Put two quarts of milk over the fire, beaten up with the yolks of four eggs and a little grated nutmeg; stir it over the fire and then gradually into the cold gooseberries ; let it stand until cold, and serve it. The eggs may be left out, and milk only may be used. Half this quantity makes a good dishful.

Stewed Prunes.

340. Take one pound of prunes, wash them in cold water, then put them into a stewpan with one quart of water, two grains of saccharin and two or three drops of cochineal ; then gradually bring to the boil, and stew gently for an hour. Serve when cold.

Lemon Jelly.

341. Soak one and a half ounces of gelatine in half a pint of water for half an hour. Put into a saucepan one pint and a half of water with the peel of one lemon and also the juice. Let it boil for a few minutes and then pour it on the gelatine ; sweeten with three grains of saccharin, return it all to the saucepan, and stir quickly into it the white and shell of one egg well-beaten. Let it gradually come to the boil, and boil for a minute ; then stand it away from the fire for two minutes, skim well and strain through a jelly-bag until clear, and then add one wineglass of brandy. When nearly cold, pour into a jelly-mould to set.

Claret Jelly.

342. Take one bottle of claret, the juice and rind of a lemon, one small pot of red-currant jelly, six grains of saccharin, one and a half ounces of isinglass, and one wineglass of brandy. Boil all together for five minutes ; strain into a mould and let it get cold ; serve with cream sauce—recipe for which as follows : Half a pint of cream, sweetened, and flavoured with vanilla and slightly whisked, poured round the jelly.

Strawberry Jelly.

343. Take a quart of fine ripe strawberries, and pour over them a pint of water that has boiled for twenty minutes, with eight grains of saccharin. The next day drain off the syrup from the strawberries without bruising them, and, to increase the fruity flavour, add a little lemon-juice. Clarify two ounces and a half of isinglass in a pint of water, and let it stand till nearly cold ; then mix it with the fruit-juice and pour into moulds.

ORANGE JELLY.

344. To the juice of eight fine sweet oranges and four Seville, well strained, add an ounce and a half of isinglass dissolved in boiling water; sweeten with four grains of saccharin, and stir it gently over the fire, but do not let it boil. Pour the jelly into moulds when nearly cold, the moulds having been previously filled with cold water.

CURRANT AND RASPBERRY JELLY.

345. Bruise in a jar two pounds of red and one pound of white currants with a pint of red raspberries; place the jar in boiling water to extract the juice. Boil three-quarters of a pint of water, two ounces of isinglass, and twelve grains of saccharin. Allow both the fruit-juice, when strained, and the sweetened isinglass to cool; then mix equal quantities, pour into shapes, and place the jelly in ice.

RUM JELLY.

346. Dissolve twelve grains of saccharin in a pint of boiling water ; mix with this two ounces of clarified isinglass or gelatine, and add the juice of a lemon and a wineglassful of fine old Jamaica rum. Pour the jelly into a damp mould, set it in ice or in a cool place till it is stiff, then turn it out and serve.

STEWED APPLES.

347. Take three or four very good American apples ; peel and core carefully ; cut into slices. Put the slices into a saucepan with a tablespoonful of water ; boil till quite tender, then beat them quite smooth with a fork, adding saccharin and lemon-juice to taste. Not the least lump or bit of core should be left.

Red Rhubarb.

348. Cut one pound of rhubarb one inch long; put into a pan with two tablespoonfuls of water and three grains of saccharin; stir on a slow fire till tender.

Compôte of Cherries.

349. Choose large, ripe, light-coloured cherries, wipe them, and leave on them about an inch of stalk, making all uniform. Put eight grains of saccharin into a saucepan with a breakfast-cupful of water, and let it boil for a minute; then put into it a pound and a half of the cherries, and simmer them for three minutes. Dish them when cold with the stalks uppermost. A tablespoonful of brandy may be added if liked.

Snow Pudding.

350. Put into half a pint of cold water half a packet of gelatine. Let it stand one hour; then add one pint of boiling water, eight grains of saccharin, and the juice of two lemons. Stir and strain, and let it stand all night; beat very stiff the whites of two eggs, and beat well into the mixture; pour into a mould.

Plain Sweet Omelette.

351. Beat the whites of two eggs to a stiff froth, add the yolks, beat again till quite smooth; put in a tablespoonful of cream, a piece of butter about the size of a walnut, and a little saccharin; put half an ounce of butter into the omelette-pan, let it just boil, and pour in the omelette; shake it till it is even all over the pan. When nicely browned on the under side, pass a red-hot salamander over the top just to set it, double up, and slide on to a hot dish, and serve at once. This omelette may be made quite plain, with only a little salt if not liked sweet. It is very light if properly made.

Custard.

352. Have the yolk of one fresh egg, beat it a little and pour over it, mixing the two well, a teacupful of thin cream;

add a little saccharin and essence of vanilla to taste; stir over
the fire in a small saucepan till it thickens, but keep it just off
the boil. A teaspoonful of brandy or rum improves it, but it
must not be added when very hot. This is nice to eat with
stewed fruit.

NOURISHING CREAM.

353. Beat the yolks of two eggs; add a little saccharin, the
rind, lightly grated, and strained juice of half a lemon or
orange. Beat the whites to a stiff froth, adding to them,
while beating, a little saccharin. Put the jar containing the
yolks in a pot of boiling water; cook gently, stirring all the
time; when it begins to thicken, stir in the whites of the eggs
until well mixed; remove it from the fire, and let it cool.
Serve in custard-glasses.

RHUBARB MOULD.

354. One quart of red rhubarb cut in pieces, put into a
saucepan with a lid; let it boil till it is a pulp. Soak half an
ounce of gelatine in cold water, pour on to it just enough boiling water to dissolve it, add it to the rhubarb with sixteen
grains of saccharin; let it boil fifteen minutes; add a few
drops of essence of lemon, butter a mould, and pour in the
rhubarb. Next day dip the mould in hot water, and turn the
rhubarb out on a glass dish.

APPLE SNOW.

355. Reduce half-a-dozen apples to a pulp, press them
through a sieve, sweeten with saccharin, and flavour them.
Take the whites of six eggs, whisk them for some minutes
with a little saccharin. Beat the pulp to a froth, then mix
the two together, and whisk them until they look like stiff
snow. Pile high in rough pieces on a glass dish, stick a sprig
of myrtle in the middle, and garnish with small pieces of
bright-coloured jelly.

APPLE SHAPE.

356. Me't a heaped tablespoonful of isinglass in a little water. Take half a pint of nicely flavoured apple-pulp, mix it well with half a pint of cream, then add the dissolved isinglass, and sweeten with saccharin to taste. Let it stand till nearly cold ; add a glass of wine or a tablespoonful of brandy, pour into a buttered mould, and keep it in a cool place until the next day.

RHUBARB FOOL.

357. Wash and, if necessary, peel the rhubarb and cut it up into small pieces. Put as much as is to be used into a jar which has a closely-fitting lid, with as much saccharin as will be required to sweeten it. Set this jar in a saucepan of boiling water, and keep it boiling until the fruit is quite soft. Rub it through a sieve with the back of a wooden spoon, and mix with the pulp as much cream as will make it of the consistency of gruel. Taste it, and, if not sufficiently sweet, add a little more saccharin. Serve cold in a glass dish.

PLUMS, COMPÔTE OF.

358. Boil half a pint of water with ten grains of saccharin ; put in a pound of plums, and let them simmer until they are tender, without being broken, if possible. Lift them out, place them on a compôte dish, and pour the syrup over them. Cream may be eaten with them.

MOULDED PEARS.

359. Peel and cut the pears into quarters ; put them into a jar, with one pint of water, eight cloves, a small piece of cinnamon, and sweeten the whole nicely with saccharin ; cover down the top of the jar, and bake the pears in a gentle oven until perfectly tender, but do not allow them to break. When done, lay the pears in a plain mould, which should be well wetted, and boil half a pint of the liquor the pears were baked in with a quarter of a pint of raisin wine, a strip of lemon-peel, the juice

of half a lemon, and one ounce of gelatine; let these ingredients boil quickly for five minutes, then strain the liquid
warm over the pears, put the mould in a cool place, and when
the jelly is firm turn it out on a glass·dish.

DAMSONS, COMPÔTE OF.

360. Take eight grains of saccharin and one pint of water;
let it simmer on the fire until the saccharin is dissolved, then
throw in the white of an egg, and take off the scum as it rises.
When the syrup has boiled fifteen minutes drop into it, one by
one, a quart of sound damsons, and simmer until soft without
breaking them. Remove them from the syrup, and boil it
again; let it cool, and pour it over the damsons; which should
have been previously arranged in a glass dish. A glass of
whipped cream is a nice accompaniment to this dish.

APRICOTS, COMPÔTE OF.

361. Take one dozen large, sound apricots; halve them,
remove the stones, and blanch the kernels. Put twelve
grains of saccharin with a pint and a half of water. Let it
boil; then put in the apricots, and let them simmer very gently
for a few minutes. Take them out, drain them, and arrange
them in a dish. When the syrup is cold, pour it over the
fruit; put half a kernel upon each piece of apricot.

BEVERAGES.

LEMONADE.

362. Put the juice of a lemon to a pint of water in which
one grain of saccharin has been dissolved; then add the white
of an egg and froth up. It may be iced.

TAMARIND WATER.

363. Boil two ounces of tamarinds with a quarter of a
pound of stoned raisins in three pints of water for an hour;
strain it, and when cold it is fit for use.

ORANGE GIN.

364. To one gallon of the strongest gin put the rinds, pared very thin, of eighteen Seville oranges and sixteen grains of saccharin. Let the gin remain on the peel in a stone jar or bottle for a month (the bottle must be kept air-tight), then bottle for use. If required it will be fit for use at once.

CHERRY BRANDY.

365. Fill wide-mouthed bottles with good Morella cherries nearly full; prick the cherries first in three or four places with a fine needle ; put into each bottle four grains of saccharin, fill up with brandy. Cork, and cover with bladder very tight. Best kept for a year.

EGGS AND BRANDY.

366. Beat up two eggs to a froth in two ounces of cold spring water ; add a little saccharin and pour in one ounce of brandy, stirring all the time.

CREAM OF TARTAR. (A Cooling Drink.)

367. Put half an ounce of cream of tartar, the juice of one lemon, and one grain of saccharin into a jug, and pour over a quart of boiling water. Cover till cold.

EGG AND SHERRY.

368. Beat up with a fork an egg till it froths; add a very small quantity of milk, and two tablespoonfuls of water; mix well, pour in a wineglassful of sherry, and serve before it gets flat.

AN AMERICAN DRINK.

369. Put the juice of a lemon to a pint of water in which one grain of saccharin has been dissolved ; then add the white of an egg and froth up. It may be iced.

CLARET CUP.

370. Take one bottle of claret, one bottle of soda-water, about half a pound of pounded ice, four grains of saccharin, a little grated nutmeg, one liqueur-glass of maraschino, and a sprig of green borage. Put all these ingredients into a silver cup, regulating the proportion of ice by the state of the weather; if very warm a larger quantity would be necessary. Hand the cup round with a clean napkin passed through one of the handles, that the edge of the cup may be wiped after each guest has partaken of the contents thereof.

EGG WINE.

371. Well beat a nice new-laid egg with a little water, and then pour over it a glass of white wine made very hot, with half a tumbler of water and a little saccharin. Stir it all the time until well mixed together; then set it over the fire until it thickens and is very hot, without coming to a boil. It must be stirred one way all the time, and when done poured into a glass, and served with a thin slice of toasted bread cut into long slices and placed on a plate crossed over each other. A little grated nutmeg may be added if the flavour is liked.

LEMONADE.

372. Grate the peel of six lemons; pour a quart of boiling water on it. Let it stand some time; then add the juice of the lemons (take care not to let the lemon pips fall into the liquid), sweeten it with one or two grains of saccharin, and run it through a jelly-bag.

WINE CAUDLE.

373. Beat with a whisk the white of one egg, and the yolks of eight; stir into it a bottle of white wine, a pint of water, the peel of a lemon, and eight grains of saccharin. Set the mixture on the fire and keep stirring it; remove it as soon as it boils. Pour the caudle into a bowl or small glasses.

FOR SUMMER DRINKS.

374. One pound of red currants bruised with some raspberries, eight grains of saccharin, added to a gallon of cold water; this is well stirred, allowed to settle, and then bottled.

Apple Water.

375. Pare and core three or four large apples, put them into a quart jug with two grains of saccharin, a few strips of very thin lemon-peel, the strained juice of half a lemon. Fill the jug with boiling water; cover it over, and leave till cold.

SAUCES FOR FISH, MEATS, VEGETABLES, AND SALADS.

Mayonnaise Sauce.

376. Take one yolk of a raw egg, salt, pepper, and a little raw mustard. Mix these together with a silver fork in a large plate; add salad oil, little by little (it will take almost any quantity, but you must be guided by taste and the quantity required). Mix by stirring one way until thick and smooth; then add vinegar enough to thin it a little. If there is any difficulty found in getting the oil to mix smoothly, add just a few drops of vinegar from time to time and keep stirring, and it will finally come right.

Lobster Sauce.

377. Break a lobster carefully, cut the meat into small pieces, beat the spawn very fine in a marble mortar with a bit of butter. Put a little melted butter on the fire, add the spawn to it, also one teaspoonful of essence of anchovy, pepper, salt, cayenne, and about half a teacup of cream; mix well, simmer a little, put in the pieces of lobster, simmer again, but do not boil, or it will curdle.

Sauce Tartare.

378. One saltspoonful of good cayenne pepper in very fine powder, half a saltspoonful of salt, a little saccharin; mix well, then add one tablespoonful of the strained juice of a lemon, two tablespoonfuls of Harvey's sauce, one teaspoonful of mushroom ketchup, and a small wineglass of port wine. Put all this into a jar, and place the jar in a pan of boiling water to heat the sauce. Very good to mix with other gravy, or to use with anything grilled.

SAUCE ESPAGNOLE.

379. Boil two eggs hard, chop up fine ; take half a cup of thin cream or good milk ; add to it the beaten raw yolk of an egg ; warm on the fire to thicken a little. Add pepper, salt, a teaspoonful of tarragon vinegar, and the chopped eggs. Heat well, and serve.

SAUCE PIQUANTE.

380. Boil together a tablespoonful of chopped onion, same of parsley, and of mushrooms, in one ounce of butter for five minutes ; then add half a pint of good stock, add salt and cayenne, and stir in last one tablespoonful of vinegar. Boil a few minutes.

CUCUMBER SAUCE.

381. Peel some cucumbers, cut up very small ; put them into a saucepan with a little broth, half a tablespoonful of vinegar, salt, cayenne, and a little essence of celery (or omit the salt and use celery salt) ; a small bit of boiled onion may be added if liked, and a bit of butter. Stew gently till tender ; rub through a sieve. Serve with any cutlets.

DUTCH SAUCE.

382. The yolks of two eggs, two tablespoonfuls of tarragon vinegar, quarter of a pint of cream, a small piece of butter, a little cayenne, and a blade of mace. To be all stirred together one way and simmered till it is the consistency of custard. Serve very hot.

HORSERADISH SAUCE.

383. Half a teaspoonful of mustard and a little salt ; work into it two tablespoonfuls of cream until quite smooth, one or

two teaspoonfuls of vinegar, and two tablespoonfuls of grated horseradish. If too thick or hot, add a little more cream. The mustard can be omitted if not liked very hot.

COLD SAUCE PIQUANTE.

384. Boil two eggs very hard, rub the yolks through a sieve, add one tablespoonful of salad oil, tarragon vinegar, chilli vinegar, and common vinegar, a little minced parsley and shallot, pepper, salt, a teacupful of cream ; stir all well and smooth together. It is better to put the vinegar in last.

SAUCE FOR COLD GAME.

385. One teaspoonful of made mustard, a very little saccharin, two teaspoonfuls of salad oil; mix smooth. Beat the yolks of two eggs, add salt, pepper, and some minced parsley ; mix well with the above, then add a teaspoonful of tarragon vinegar and half a glass of white wine.

SALAD MIXTURE.

386. Two yolks of hard-boiled eggs, rubbed very smooth, with a good pinch of salt, a little saccharin, a teaspoonful of raw mustard ; mix all well. Next, three tablespoonfuls of salad oil worked in by degrees, four tablespoonfuls of cream ; lastly, one to two tablespoonfuls of vinegar. More cream can be added before the vinegar if too thick or if it is too hot of mustard.

PUDDING SAUCE.

387. The beaten yolk of an egg, mixed with a little saccharin and half a teacupful of cream. Stir over the fire until it thickens, then add half a wineglass of wine ; warm it, but do not let it boil.

LEMON SAUCE.

388. Cut the rind of a lemon very thin ; boil it for three minutes in a teacup of water ; stir in the juice of the lemon, strained, and add a little saccharin and a few drops of brandy.

SAUCE PIQUANTE.

389. Half a teacup of gravy, one tablespoonful of French mustard, two of Worcester sauce, two of port wine, a little saccharin, one teaspoonful of shallot and chilli vinegar. Warm gently over a slow fire, and serve over warmed game; or pieces of meat or game can be warmed in it.

BROWN ONION SAUCE.

390. Slice some onions, about five; brown in a stewpan with butter; add half a pint of good stock, and stew till tender.

TOMATO SAUCE.

391. Melt in a stewpan a dozen or two ripe tomatoes (which, before putting into the stewpan, cut in two and squeeze the juice and the seeds out). Then put two shallots, one onion, a clove, a little thyme, a bay-leaf, a few leaves of mace, and when melted rub them through a tamis. Mix a few spoonfuls of good espagnole and a little salt and pepper with this purée. Boil it for twenty minutes, and serve.

MINT SAUCE.

392. Wash and free from grit three tablespoonfuls of young green mint, chop exceedingly fine and put it in a sauce-tureen with a teacupful of vinegar, and sweeten according to taste with saccharin. Mint sauce should be allowed to stand an hour or two before being used.

TARTAR SAUCE.

393. Rub the hard-boiled yolks of three eggs to a powder; add a saltspoonful of mustard, half a saltspoonful of salt, half a grain of cayenne, and the beaten yolk of one egg; stir in drop by drop four tablespoonfuls of lucca oil, two table-spoonfuls of tarragon vinegar, and one tablespoonful of French vinegar; continue to stir till the sauce becomes thick;

9

chop quite fine one shallot, a piece of garlic as big as a pea, and one small gherkin; stir these into the sauce, and serve cold.

Royal Sauce for Fish.

394. Beat two raw yolks of eggs with two ounces of fresh butter; add gradually a teaspoonful of elder vinegar, a teaspoonful of tarragon vinegar, a teaspoonful of soy, a pinch of cayenne, and a very small quantity of nutmeg. Pour the mixture into an earthen jar; set this in a small saucepan of boiling water, and keep it boiling, stirring briskly all the time, until the sauce begins to thicken and presents a rich, smooth appearance; be careful that it does not curdle, which it will quickly do if not taken from the fire as soon as it is smooth and thick. Time, about ten minutes.

Italian Sauce.

395. Put the following ingredients into a stewpan: Two spoonfuls of chopped mushrooms, one of parsley, half a shallot, the same of bay-leaf; add pepper and salt to taste. Stew them gently with just enough espagnole sauce to moisten them, and thin to a proper consistency with good strong broth.

Brown Butter Sauce for Fish.

396. Dissolve two ounces of butter in a saucepan, and stir it till it is brown without burning. Add two tablespoonfuls of tarragon vinegar, four tablespoonfuls of good brown sauce, a tablespoonful of Harvey, a teaspoonful of bruised capers, and half a teaspoonful of anchovy. Stir the same over the fire till it boils, and serve.

Spinach Sauce for Boiled Fowls, etc.

397. Wash the spinach in two or three waters, pick the leaves from the stalks, drain it, and stew it with as much water only as will keep it from burning. Squeeze the moisture

from it, and beat it with a wooden spoon till smooth. Dissolve
a slice of fresh butter in a saucepan, put in the spinach, and
stir it till it is quite hot and dry. Add pepper and salt, and as
much boiling milk as will make the same of the consistency of
thick cream.

MUSHROOM SAUCE.

398. Button or flap mushrooms may be used for this sauce.
They should be rinsed in cold water, drained, and dried in a
soft cloth, and if flap ones cut into pieces. Simmer the mushrooms, without stalks, in half a pint of beef gravy ; add a little
mushroom ketchup and an ounce of butter. If liked, flavour
with lemon peel, and squeeze in some of the juice before
serving.

APPLE SAUCE.

399. Pare, core, and slice four or five apples ; place them in
a saucepan with only just enough water to keep them from
burning. Let them simmer gently, stirring frequently, over a
slow fire, until they are reduced to a pulp ; turn them into a
bowl, and beat them well ; sweeten with saccharin according to
taste, and add the squeeze of a lemon, and a small piece of
butter.

GOOSEBERRY SAUCE.

400. Cut the tops and stalks from half a pint of green
gooseberries; boil them until tender, press them through a
sieve, and mix them with a little butter. Various seasonings
may be used for this sauce, such as grated ginger or grated
lemon-rind, grated nutmeg, a little saccharin, or cayenne
pepper ; a wineglassful of sorrel or spinach-juice is a decided
improvement.

SAUCE FOR MUTTON CHOPS.

401. Take three tablespoonfuls of gravy, two of wine, two
of Worcester sauce or ketchup, salt, pepper, and a teaspoonful
of shallot vinegar ; stir till hot; pour over the chops.

LEMON SAUCE.

402. Cut the rind of a lemon very thin, boil it for three minutes in a teacup of water, stir in the juice of the lemon, strained; add a little saccharin and a few drops of brandy.

PUDDING SAUCE.

403. The beaten yolk of an egg, mixed with a little saccharin and half a teacupful of cream. Stir over the fire until it thickens, then add half a wineglass of wine; warm it, but don't let it boil.

INDEX

The figures refer to the number of the paragraphs.

THE END.

www.ingramcontent.com/pod-product-compliance
Lightning Source LLC
Chambersburg PA
CBHW021817190326
41518CB00007B/633